Philosophy of Biology

Dimensions of Philosophy Series
Norman Daniels and Keith Lehrer, Editors

Philosophy of Biology, Elliott Sober

Metaphysics, Peter van Inwagen

Philosophy of Physics, Lawrence Sklar

Theory of Knowledge, Keith Lehrer

Philosophy of Law: An Introduction to Jurisprudence,
Revised Edition, Jeffrie G. Murphy and Jules L. Coleman

Philosophy of Social Science, Alexander Rosenberg

Introduction to Marx and Engels: A Critical Reconstruction,
Richard Schmitt

FORTHCOMING

Philosophy of Science, Clark Glymour

Philosophy of Mind, Jaegwon Kim

Contemporary Continental Philosophy, Bernd Magnus

Political Philosophy, Jean Hampton

Normative Ethics, Shelly Kagan

Philosophy of Religion, Thomas V. Morris

Philosophy of Education, Nel Noddings

Philosophy of Language, Stephen Neale

Also by Elliott Sober

Simplicity

The Nature of Selection: Evolutionary Theory in Philosophical Focus

Reconstructing the Past: Parsimony, Evolution, and Inference

Core Questions in Philosophy

Reconstructing Marxism: Explanation and the Theory of History
(with Erik Wright and Andrew Levine)

Philosophy of Biology

Elliott Sober

University of Wisconsin–Madison

Westview Press

Boulder • San Francisco

Dimensions of Philosophy Series

Published in 1993 in the United States of America by Westview Press, Inc., 5500 Central Avenue, Boulder, Colorado 80301-2877

Library of Congress Cataloging-in-Publication Data
Sober, Elliott.
Philosophy of biology / Elliott Sober.
p. cm. — (Dimensions of philosophy series)
Includes bibliographical references (p.) and index.
ISBN 0-8133-0785-6. — ISBN 0-8133-0824-0 (pbk.)
1. Evolution (Biology)—Philosophy. 2. Creationism. 3. Natural selection—Philosophy. 4. Evolution (Biology)—Religious aspects—Christianity. I. Title. II. Series.
QH371.S62 1993
575′.001—dc20 92-37484
CIP

Printed and bound in the United States of America

The paper used in this publication meets the requirements of the American National Standard for Permanence of Paper for Printed Library Materials Z39.48-1984.

10 9 8 7 6 5 4

for Sam

Contents

Boxes

Acknowledgments

Robert Boyd, Robert Brandon, David Hull, Robert Jeanne, Philip Kitcher, John Maynard Smith, Robert O'Hara, Steven Hecht Orzack, Peter Richerson, Louise Robbins, Robert Rossi, Michael Ruse, Kim Sterelny, and David Sloan Wilson gave me plenty of useful advice on how earlier versions of this book could be improved. I am very grateful to them for their help.

I have reprinted material in Chapter 5 from an article I coauthored with Steven Hecht Orzack, "Optimality Models and the Long-run Test of Adaptationism" (*American Naturalist,* forthcoming). Chapter 4 contains passages from my essay "The Evolution of Altruism: Correlation, Cost, and Benefit" (*Biology and Philosophy,* 1992, 7: 177–187, copyright © 1992 by Kluwer Academic Publishers; reprinted by permission of Kluwer Academic Publishers). In Chapter 7, I have used portions of my essays "Models of Cultural Evolution" (in P. Griffiths, ed., *Trees of Life,* 1991, pp. 17–39, copyright © 1991 by Kluwer Academic Publishers; reprinted by permission of Kluwer Academic Publishers) and "When Biology and Culture Conflict" (in H. Rolston III, ed., *Biology, Ethics, and the Origins of Life,* forthcoming). I thank my coauthor and the editors and publishers for permitting me to use these passages.

Elliott Sober

Introduction

This book concentrates on philosophical problems raised by the theory of evolution. Chapter 1 describes some of the main features of that theory. What *is* evolution? What are the principal elements of the theory that Charles Darwin proposed and that subsequent biology has elaborated? How is evolutionary biology divided into subdisciplines? How is evolutionary theory related to the rest of biology and to the subject matter of physics?

After this preliminary chapter (some of whose themes are taken up later), the book is divided into three unequal parts. The first concerns *the threat from without*. Creationists have challenged the theory of descent with modification and have defended the idea that species were separately created by God. My treatment of creationism is not a detailed empirical defense of evolutionary theory. Rather, I am interested in the logic of both the creationist argument and Darwin's hypothesis of common ancestry. I also discuss an issue of general significance in the philosophy of science: What makes a hypothesis scientific? Creationists have used answers to this question as clubs against evolutionary theory; evolutionists have reciprocated by attempting to show that "scientific creationism" is a contradiction in terms. In light of all this combat, the difference between science and nonscience is worth examining with more care.

The second and largest portion of the book concerns philosophical issues that are internal to evolutionary biology: The debates I address here involve *turmoil within*. Chapter 3 is a preliminary to this set of biological issues. The theory of natural selection is fundamental to evolutionary biology, and the concept of fitness is central to that theory. Therefore, we must understand what fitness is. We also must see how it makes use of the concept of probability. And we must examine why the concept of fitness is useful in constructing evolutionary explanations.

Chapter 4 explores a fascinating debate that has enlivened evolutionary theory ever since Darwin. It centers on the issue of *the units of selection*. Does natural selection cause characteristics to evolve because they are good for the species, good for the individual organism, or good for the genes? An important part of this problem concerns the issue of *evolutionary altruism*. An altruistic characteristic is deleterious to the individual possessing it, though beneficial to the group in which it occurs. Is altruism an outcome of the evolutionary process, or does evolution give rise to selfishness and nothing else?

Chapter 5 turns to another debate that currently occupies biologists. Many evolutionists believe that natural selection is overwhelmingly the most important cause of the diversity we observe in the living world. Others have criticized this emphasis on selection and have argued that *adaptationists* accept this guiding idea uncritically. In Chapter 5, I try to clarify what this debate is about. I also discuss how adaptive explanations should be tested.

Chapter 6 moves away from the process of natural selection and focuses on the patterns of similarity and difference that evolution produces. How are organisms to be grouped into species? How should species be grouped into higher taxa? Here, I consider the part of evolutionary biology called *systematics*. Evolutionary theory says that species are genealogically related to each other. How is the system of ancestor/descendant relationships exhibited by the tree of life to be inferred?

If the first part of this book concerns *the threat from without* and the second part describes *turmoil within*, the third may be said to describe *the urge to expand*. Chapter 7 analyzes a variety of philosophical issues raised by the research program called *sociobiology*. I say that this chapter concerns the urge to expand because sociobiology is often thought to be an imperialistic research program; it aims to expropriate phenomena from the social sciences and show that they can be given biological explanations.

Sociobiologists consider an organism's behavior, no less than the shape of its bones or the chemistry of its blood, to be a topic for evolutionary explanation. Since human beings are part of the evolutionary process, sociobiologists see no reason to exempt human behavior from evolutionary treatment. How much of human behavior can be understood from an evolutionary perspective? Perhaps the fact that we have minds and participate in a culture makes it inappropriate to apply evolutionary explanations to our species. On the other hand, perhaps exempting ourselves from the subject matter of evolutionary theory is just wishful thinking, a reflection of the naive self-love that leads human beings to think that they are outside of, rather than a part of, nature. Sociobiology has ignited a passionate debate. It touches directly on the question of what it means to be human.

These, then, are the main biological subjects I discuss. Each is the occasion for examining a variety of philosophical issues. Vitalism and materialism get a hearing. Reductionism and its antithesis also come in for discussion, as do likelihood inferences and Karl Popper's falsifiability criterion. I will examine the problem of interpreting the probability concept and the meaning of randomness, of correlation, and of Simpson's paradox. I'll also discuss the role of teleological concepts in science. And discussing the species problem will provide a context for addressing the larger issue of essentialism. The role of Ockham's razor in scientific inference will be analyzed as well.

As this ragtag list suggests, I've organized this book mainly around biological concepts and problems, not around philosophical *isms*. I shudder at the thought of trying to organize all of the philosophy of biology in terms of a contest between warring philosophical schools. I am not inclined to see biology as a test case for positivism or for reductionism or for scientific realism. This is not because I find these philosophical *isms* uninteresting but because the organizing principle I prefer is to have the philosophy of biology grow out of the biology.

My preeminent focus on evolutionary theory deserves a comment, if not an apology. There is more to biology than the theory of evolution, and there is more to the philosophy of biology than the set of problems I have chosen to examine here. For example, much of the large body of literature on reductionism has considered the relationship of Mendelian genetics to molecular biology. And the philosophy of medicine and environmental ethics are burgeoning fields. In discussing other biological areas only briefly, I do not mean to imply that they are unworthy of philosophical attention. My selection of topics is the result of my interests plus the fact that this book is supposed to be reasonably short. For me, evolutionary biology is the center of gravity for both the science of biology and for the philosophy of that science. The philosophy of biology does not end with evolutionary issues, but that is where I think it begins. I believe that a number of the points I'll make about the theory of evolution generalize to other areas of biology and to some of the rest of science besides. Readers must judge for themselves whether this is so.

1

What Is Evolutionary Theory?

1.1 What Is Evolution?

We talk of stars evolving from red giants to white dwarfs. We speak of political systems evolving toward or away from democracy. In ordinary parlance, "evolution" means *change*.

If evolution is understood in this way, then the theory of evolution should provide a global account of cosmic change. Laws must be stated in which the trajectories of stars, of societies, and of everything else are encapsulated within a single framework. Indeed, this is what Herbert Spencer (1820–1903) attempted to do. Whereas Charles Darwin (1809–1882) proposed a theory about how *life* evolves, Spencer thought he could generalize Darwin's insights and state principles that govern how *everything* evolves.

Although the allure of a unified theory of everything is undeniable, it is important to realize that evolutionary biology has much more modest pretensions. Evolutionary biologists use the term "evolution" with a narrower meaning. One standard definition says that evolution occurs precisely when there is change in the gene frequencies found in a population. When a new gene is introduced or an old one disappears or when the mix of genes changes, the population is said to have evolved. According to this usage, stars do not evolve. And if political institutions change because people change their minds, not their genes, then political evolution is not evolution in the biologist's proprietary sense.

Biologists usually compute gene frequencies by *head counting*. Suppose two lizards are sitting on a rock; they are genetically different because one possesses gene *A* and the other possesses gene *B*. If one grows fat while the other grows thin, the number of cells containing *A* increases and the number of cells containing *B* declines. However, the gene frequencies, computed *per capita,* remain the same. The growth of organisms (their *ontogeny*) is not the same thing as the evolution of a population (Lewontin 1978).

The idea that change in gene frequency is the touchstone of evolution does not mean that evolutionists are interested only in genes. Evolutionary biologists try to figure out, for example, why the several species in the horse lineage increased in height. They also seek to explain why cockroaches have become more resistant to DDT. These are changes in the *phenotypes* of organisms—in their morphology, physiology, and behavior.

When a population increases its average height, this may or may not be due to a genetic change. Children may be taller than their parents simply because the quality of nutrition has improved, not because the two generations are genetically different. However, in the case of the horse lineage, biologists believe that the increasing height of successive species does reflect a change in their genetic endowment. The definition of evolution as change in gene frequency will count some cases of height increase—but not others—as instances of evolution. This definition does not deny that phenotypic change can count as evolution. What it rejects is change that is "merely" phenotypic.

Another worry is that the definition of evolution as a change of gene frequencies ignores the fact that evolution involves the origin of new species and the disappearance of old ones. Evolutionists use the term *microevolution* to describe the changes that take place within a persisting species. *Macroevolution* is reserved for the births and deaths of species and higher taxa. Does the definition of evolution as change in gene frequency mean that macroevolution is not evolution? This is not a consequence of the definition, as long as daughter species differ genetically from their parents. If speciation—the process by which new species come into being—entails change in gene frequency, then speciation counts as evolution as far as this definition is concerned.

To further explore this definition of evolution, we need to review some elementary biology. Genes are found in chromosomes, which, in turn, are found in the nuclei of cells. It is a simplification, though a useful one for getting started, to think of the genes in a chromosome as arranged like beads on a string. Some species—including humans—have chromosomes in pairs. Such species are said to be *diploid*. Others have their chromosomes as singletons (*haploid*) or in threes (*triploid*) or fours (*tetraploid*). A species also may be characterized by how many chromosomes the organisms in it possess.

If we consider a pair of chromosomes in a diploid organism, we can ask what gene occurs on each of the two chromosomes at a given location (a *locus*). If there is only one form that the gene can take, then all members of the species are identical at that locus. However, if more than one form (*allele*) of the gene occurs, then the organisms will differ from each other at that locus.

Suppose there are two alleles that a diploid organism may have at a given location, which I'll call the *A*-locus. These alleles I'll call *A* ("big *A*") and *a* ("little *a*"). Each organism will either have two copies of *A* or two copies of *a* or one copy of each. The *genotype* of the organism at that locus is the pair of genes it possesses there. *AA* and *aa* organisms are termed *homozygotes; Aa* individuals are called *heterozygotes*.

Now I come to sex. This is a common but by no means universal mode of reproduction. A diploid organism forms *gametes*, which contain just one of the two chromosomes that occur in each chromosomal pair: The gametes are haploid. The process by which diploid parents produce haploid gametes is called *meiosis*. An individual who is heterozygous at the *A*-locus typically will have 50 percent *A* gametes and 50 percent *a* gametes (though not always—see Section 4.5). The nonsex cells (somatic cells) in an individual are genetically identical with each other (ignoring for the moment the infrequent occurrence of mutations), but the gametes that an individual produces may be immensely different because the individual is heterozygous at various loci. Diploid parents produce haploid gametes, which come together in reproduction to form a diploid offspring.

If I describe the genotypes of all the males and females in a population, can you figure out what the genotypes will be of the offspring they produce? The answer is no. You need to know who mates with whom. If a mother and father are both *AA* (or both *aa*), their offspring will all be *AA* (or *aa*). But when heterozygotes mate with heterozygotes (or with homozygotes), the offspring may differ from each other.

Mating is said to be *random* within a population if each female is as likely to mate with one male as with any other (and *vice versa*). Mating is *assortative*, on the other hand, if similar organisms tend to choose each other as mates. I now want to describe how assortative mating provides a counterexample to the claim that evolution occurs precisely when there is change in gene frequencies.

Suppose that each organism mates only with organisms that have the same genotype at the *A*-locus. This means that there are only three kinds of crosses in the population, not six. These are $AA \times AA$, $aa \times aa$, and $Aa \times Aa$. What are the evolutionary consequences of this pattern of mating?

Consider a concrete example. Suppose the process begins with 400 individuals, of which 100 are *AA*, 200 are *Aa*, and 100 are *aa*. Notice that there are 800 alleles in the population at the locus in question (2 per individual times 400 individuals). Notice further that there are 400 copies of *A* (200 in the homozygotes and 200 in the heterozygotes) and 400 copies of *a*. So, initially, the gene frequencies are 50 percent *A* and 50 percent *a*.

Suppose these 400 individuals pair up, mate, and die, with each mating pair producing 2 offspring. In the next generation, there will be 400 individuals. The following table describes the productivities of the mating pairs:

Parental pairs		*Offspring*
50 $AA \times AA$	produce	100 *AA*
100 $Aa \times Aa$	produce	50 *AA* 100 *Aa* 50 *aa*
50 $aa \times aa$	produce	100 *aa*

If you don't understand how I calculated the numbers of different offspring in the heterozygote mating, don't worry. The present point is simply that not all the offspring of such matings are heterozygotes.

Let's compare the frequencies of the three genotypes before and after reproduction. Before, the ratios are 1/4, 1/2, 1/4. After, they are 3/8, 1/4, 3/8. The frequency of heterozygosity has declined.

What has happened to the gene frequencies in this process? Before reproduction, *A* and *a* were each 50 percent. Afterwards, the same is true. There are 800 alleles present in the 400 offspring—400 copies of *A* (300 in homozygotes and 100 in heterozygotes) and 400 copies of *a*. The frequencies of genotypes have changed, but the gene frequencies have not.

In this example, the population begins at precisely 50 percent *A* and 50 percent *a*, and the assortative pattern is perfect—like *always* mates with like. However, neither of these details is crucial to the pattern that emerges. No matter where the gene frequencies begin and no matter how biased the pattern of positive association, assortative mating causes the frequency of heterozygosity to decline though gene frequencies remain unchanged.

Is the process generated by assortative mating an evolutionary one? It is standard fare in evolution texts and journals. To exclude it from the subject matter of evolutionary theory would be a groundless stipulation. I conclude that evolution does not require change in gene frequency.

Genes are important in the evolutionary process. But the gene *frequency* in a population is only one mathematical description of that population. For example, it fails to describe the frequencies of gene *combinations* (e.g., genotypes). The mistake in the definition of evolution as change in gene frequency comes from thinking that this single mathematical description always reflects whether an evolutionary change has taken place.

Genes are related to genotypes as parts are related to wholes: Genotypes are *pairs* of genes. This may lead one to expect that by saying what is true of the genes, one thereby settles what is true of the genotypes. After all, if I tell you what is going on in each cell of your body, doesn't that settle the question of what is going on in your body as a whole? This expectation is radically untrue when the properties in question are frequencies. Describing the frequencies of genes does *not* determine what the genotype frequencies are. For this reason, genotype frequencies can change while gene frequencies remain constant.

A second question about the definition of evolution as change in gene frequency is worth considering. I said earlier that genes are found in chromosomes, which are located in the nuclei of cells. However, it has been known for some time that there are bodies outside the nuclei (in the *cytoplasm*) that can provide a mechanism of inheritance (Whitehouse 1973). *Mitochondria* influence various phenotypic traits, and the DNA they contain is inherited. If a population changes its mitochondrial characters while its chromosomal features remain the same, is this an instance of evolution? Perhaps we should stretch the concept of the gene to include extrachromosomal factors. This would allow us to retain the definition of evolution as change in gene frequency, though, of course, it raises interesting questions about what we mean by a "gene" (Kitcher 1982b).

Another feature of the definition of evolution as change in gene frequency is that it does not count as evolution a mere change in the *numbers* of organisms a species contains. If a species expands or contracts its range, this is of great *ecological* significance, and a historian of that species will want to describe such changes in habitat. But if this change leaves gene frequencies unchanged, should it be excluded from the category of evolution? I won't try to answer this question. The point, again, is that *change in gene frequency* covers one type of change but fails to include others.

A final limitation in the definition of evolution as change in gene frequency is noteworthy. The genetic system itself is a product of evolution. Hence, an evolutionary process was underway before genes even existed. This objection to the standard definition is perhaps the most serious one, because it is difficult to see what better definition could be constructed in response.

The term "evolution" denotes the subject matter of an extremely variegated discipline, whose subfields differ in their aims, methods, and results. In addition, evolutionary biology is a developing entity, extending (and contracting) its boundaries in several directions at once. We should not be surprised that it is hard to delimit the subject matter of such a discipline with absolute precision. In Section 6.1, I will discuss the idea that a biological species cannot be defined by specifying necessary and sufficient conditions that the organism in it must fulfill. The same idea applies to a scientific discipline; it also evolves, so we sometimes will be unclear as to whether a given phenomenon is within its purview.

It should not disturb us if "evolution" cannot be defined precisely; the integrity of a subject is not thrown in doubt if the phenomena it addresses cannot be isolated with absolute clarity. Defining evolution is a useful first step in understanding what evolutionary biology is about; beyond that, it is a mistake to require more precision than is possible or necessary.

1.2 The Place of Evolutionary Theory in Biology

Theodosius Dobzhansky (1973) once said that "nothing in biology makes sense except in the light of evolution." It is perhaps not surprising that the man who said this was himself an evolutionary biologist. What is the relationship of evolutionary theory to the rest of biology?

Many areas in biology focus on nonevolutionary questions. Molecular biology and biochemistry, for example, have experienced enormous growth since James Watson and Francis Crick discovered the physical structure of DNA in 1953. They did not address the question of *why* DNA is the physical basis of the genetic code. This is an evolutionary question, but it is not the one they posed in their studies. Ecology is another area that often proceeds without engaging evolutionary issues. An ecologist might seek to describe the food chain (or web) that exists in a community of coexisting species (in a valley, say). By discovering who eats whom, the ecologist will understand how energy flows through the ecological system. Although gene frequencies may be changing within the species that the ecologist describes, this is not the fundamental focus of his or her investigation.

There is no need to multiply examples beyond necessity. If so much of biology proceeds without attending to evolutionary questions, why should we think that evolutionary theory is central to the rest of biology?

Box 1.1 Definitions

Philosophers often try to provide *definitions* of concepts (e.g., of *knowledge*, *justice*, and *freedom*). However, one view of definitions suggests that this activity is silly. This involves the idea that definitions are *stipulations*. We arbitrarily decide what meaning we will assign to a word. Thus, "evolution" can be defined any way we please. On this view, there is no such thing as a *mistaken* definition. As Lewis Carroll's Humpty Dumpty once observed, we are the masters of our words, not *vice versa*.

If stipulative definitions were the only kind of definition, it *would* be silly to argue about whether a definition is "really" correct. But there are two other kinds to consider.

A *descriptive* definition seeks to record the way a term is used within a given speech community. Descriptive definitions can be mistaken. This sort of definition is usually of more interest to lexicographers than to philosophers.

An *explicative* definition aims not only to capture the way a concept is used but also to make the concept clearer and more precise. If a concept is used in a vague or contradictory way, an explicative definition will depart from ordinary usage. This type of definition, which in a sense falls between stipulation and description, is often what philosophers try to formulate.

We can locate evolutionary theory in the larger scheme of things by considering Ernst Mayr's (1961) distinction between *proximate explanation* and *ultimate explanation*. Consider the question "Why do ivy plants grow toward the sunlight?" This question is ambiguous. It could be asking us to describe the mechanisms present inside each plant that allow the plant to engage in phototropism. This is a problem to be solved by the plant physiologist. Alternatively, the question could be taken to ask why ivy plants (or their ancestors) evolved the capacity to seek the light.

The plant physiologist sees a plant growing toward the light and connects that effect with a cause that exists within the organism's own lifetime. The evolutionist sees the same phenomenon but finds an explanation in the distant past. The plant physiologist tries to describe a (relatively) proximal ontogenetic cause, whereas the evolutionist aims to formulate a more distal (or "ultimate") phylogenetic explanation.

This distinction does not mean that evolutionary theory has the best or deepest answer to every question in biology. "How do the mechanisms inside a plant allow it to seek the light?" is not an evolutionary question at all. Rather, evolutionary questions can be raised about any biological phenomenon. Evolutionary theory is important because evolution is always in the background.

Evolutionary theory is related to the rest of biology the way the study of history is related to much of the social sciences. Economists and sociologists are interested in describing how a given society currently works. For example, they might study the post–World War II United States. Social scientists will show how causes and effects are related within the society. But certain facts about that society—for instance, its configuration right after World War II—

will be taken as given. The historian focuses on these elements and traces them further into the past.

Different social sciences often describe their objects on different time scales. Individual psychology connects causes and effects that exist within an organism's own lifetime. Sociology and economics encompass longer reaches of time. And history often works within an even larger time frame. This intellectual division of labor is not entirely dissimilar to that found among physiology, ecology, and evolutionary theory.

So Dobzhansky's remark about the centrality of evolutionary theory to the rest of biology is a special case of a more general idea. Nothing can be understood *ahistorically*. Of course, what this really means is that nothing can be understood *completely* without attending to its history. Molecular biology has provided us with considerable understanding of the DNA molecule, and ecology allows us to understand something about how the food web in a given community is structured. By ignoring evolution, these disciplines do not ensure that their inquiries will be *fruitless*. Ignoring evolution simply means that the explanations will be *incomplete*.

Does Dobzhansky's idea identify an asymmetry between evolutionary theory and other parts of biology? Granted, nothing in biology can be understood *completely* without attending to evolution. But the same can be said of molecular biology and of ecology: No biological phenomenon can be understood *completely* without inputs from these two disciplines. For example, a complete understanding of phototropism will require information from molecular biology, from ecology, and from evolutionary theory.

I leave it to the reader to consider whether more can be said about evolutionary theory's centrality than the modest point identified here. Evolution matters because history matters. Evolutionary theory is the most historical subject in the biological sciences, in the sense that its problems possess the longest time scales.

1.3 Pattern and Process

Current evolutionary theory traces back to Darwin. This does not mean that current theorists agree with Darwin in every detail. Many biologists think of themselves as elaborating and refining the Darwinian paradigm. Others dissent from it and try to strike out along new paths. But for disciples and dissenters as well, Darwinism is where one begins, even though it may not be where one ends.

Darwin's theory of evolution contains two big ideas, neither of them totally original with him. What was original was their combination and application. The first ingredient is the idea of a *tree of life*. According to this idea, the different species that now populate the earth have common ancestors—human beings and chimps, for instance, derive from a common ancestor. The strong form of this idea is that there is a *single* tree of terrestrial life. That is, for any two current species, there is a species that is their common ancestor—not only are we related to chimps, we also are related to cattle, to crows, and to crocuses. Weaker forms of the tree of life hypothesis also are possible.

Box 1.2 How Versus Why

It might be suggested that physiology tells us *how* organisms manage to do what they do but that evolutionary theory tells us *why* they behave as they do (see Alcock 1989 for discussion). The first clarification needed here is that *how* and *why* are not mutually exclusive. To say how ivy plants manage to grow toward the light is to describe structures that *cause* the plants to do so. The presence of these internal structures explains why the plants grow toward the light. Physiologists answer why-questions just as much as evolutionists do.

Nonetheless, there is a division of labor between the physiologist and the evolutionist. Each answers one but not the other of the following two questions: (1) What mechanisms inside ivy plants cause them to grow toward the light? (2) Why do ivy plants contain mechanisms that cause them to grow toward the light? Question (1) calls for details about *structure;* question (2) naturally leads one to consider issues pertaining to *function* (Section 3.7).

In a causal chain from *A* to *B* to *C*, *B* is a proximal cause of *C*, while *A* is a more distal cause of *C*. In a sense, *A* explains more than *B* does since *A* explains both *B* and *C*, while *B* explains just *C*. Can this difference between proximal and distal causes be used to argue that evolutionary biology is the deepest and most fundamental part of biology?

The idea of a tree of life obviously entails the idea of *evolution*. If human beings and chimps have a common ancestor, then there must have been change in the lineages leading from that ancestor to its descendants. But the tree of life hypothesis says more than just that evolution has occurred.

To see where this extra ingredient comes in, consider a quite different conception of evolution, one developed by Jean Baptiste Lamarck (1744–1829). Lamarck (1809) thought that living things contain within themselves an inherent tendency to increase in complexity. He believed that simple life forms spring from nonliving material, and from the simplest forms, more complicated species are descended. The lineage that we belong to is the oldest, Lamarck thought, because human beings are the most complicated of creatures. Modern earthworms belong to a younger lineage since they are relatively simple. And according to Lamarck's theory, present-day human beings are *not* related to present-day earthworms. This idea is quite consistent with his belief that present-day human beings are descended from earthworms that lived long ago.

Darwin thought that present and past species form a *single tree*. Lamarck denied this. Both offered theories of evolution; both endorsed the idea of descent with modification. But they differed with respect to the pattern of ancestor/descendant relationships that obtains among living things. The idea of a single tree of life is a feature of current evolutionary theory; I'll discuss the evidence for this idea in Chapter 2.

If we described the tree of life in some detail, we would say which species are descended from which others and when new characteristics originated and old ones disappeared. What is left for evolutionary theory to do, once these facts about life's pattern are described? One task that remains is to address the question of *why*. If a new characteristic evolved in a lineage, why did it do so? And if a new species comes into existence or an old one exits from the scene, again the question is why that event occurred. Answers to such questions involve theories about the *process* of evolution. As we move from the root to the tips of the tree of life, we see speciation events, extinctions, and new characteristics evolving. What processes occur in the tree's branches that explain these occurrences?

Darwin's answer to this question about process constituted the second ingredient in his theory of evolution. This is the idea of *natural selection*. The idea is simple. Suppose the organisms in a population differ in their abilities to survive or reproduce. This difference may have a variety of causes. Let's consider a concrete example—a herd of zebras in which there is variation in running speed. Suppose that faster zebras are better able to survive because they are better able to evade predators. Let us further suppose that running speed is inherited; offspring take after their parents. What will happen to the average speed in the herd, given these two facts? The Darwinian idea is that natural selection will favor faster zebras over slower ones, and so, gradually, the average running speed in the herd will increase. This may take many generations if the differences in speed are slight. But small advantages, accumulated over a large number of generations, can add up.

There are three basic constituents in the process of evolution by natural selection. First, there must be *variation* in the objects considered; if all the zebras ran at the same speed, there would be no variation on which selection could act. Second, the variation must entail variation in *fitness;* if running speed made no difference to survival or reproduction, then natural selection would not favor fast zebras over slow ones. Third, the characteristics must be *inherited;* if the offspring of fast parents weren't faster than the offspring of slow parents, the fact that fast zebras survive better than slow ones would not change the composition of the population in the next generation. In short, evolution by natural selection requires that there be *heritable variation in fitness* (Lewontin 1970).

The idea of variation in fitness is easy enough to grasp. But the third ingredient—heritability—requires more explanation. Two ideas need to be explored. First, zebras reproduce sexually. What does it mean to say that running speed is heritable if each offspring has two parents who may themselves differ in running speed? Second, we need to see why the absence of heritability can prevent the population from evolving, even when selection favors fast zebras over slow ones.

The modern idea of heritability (a statistical concept) can be understood by examining Figure 1.1. Suppose we take the running speed of the male and female in a parental pair and average them; this is called the "midparent speed." We then record the running speed of each of their offspring. Each offspring can be represented as a data point—the x axis records the average running speed of its parents, and the y axis represents its own running speed.

Notice that a given parental pair produces offspring that run at different speeds and that two offspring may have the same running speed, even

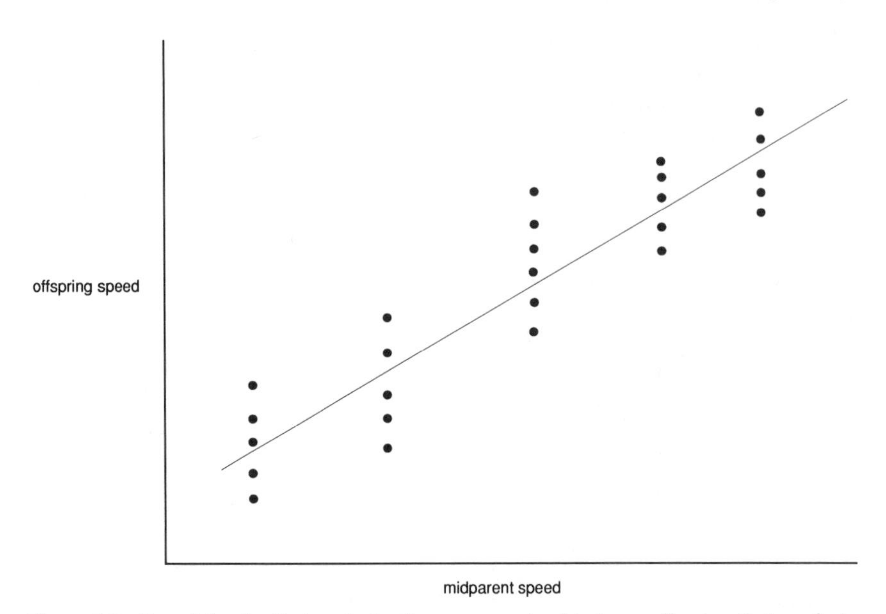

Figure 1.1 Parental pairs that are faster than average tend to have offspring that are faster than average. Running speed is *heritable;* the line drawn through the data points has a positive slope.

though they came from different parents. However, Figure 1.1 shows that *on average, faster parents tend to have faster offspring*. The line drawn through these data points represents this fact about the averages. When we say that evolution by natural selection requires heritability, this doesn't mean that offspring must *exactly* resemble their parents. In fact, this almost never happens when organisms reproduce sexually. What is required is just that offspring "tend" to resemble their parents. This claim about tendency is represented by the fact that the line in Figure 1.1 runs uphill.

Suppose there were zero heritability in running speed. Parents differ in their running speed, and faster parents tend to have more offspring than slower parents. But, on average, fast parents produce the same mix of fast and slow offspring that slow parents produce. If so, the line in Figure 1.1 will have zero slope. What will happen in this case? Natural selection will permit fast organisms to survive to reproductive age more successfully than slow organisms. But the higher representation of fast organisms at the adult stage will have no effect on the composition of the population in the next generation. Fast parents will produce the same mix of slow and fast offspring that slow parents produce. The result is that the next generation will fail to differ from the one before. Evolution by natural selection requires that the evolving trait be heritable.

Notice that this description of heritability makes no mention of genes. The description involves the relationship between parental and offspring phenotypes. How, then, do genes enter into the idea of heritable variation in fitness? If we ask *why* offspring tend to resemble their parents, the explanation may be that offspring and parents are genetically similar. Fast parents are fast and slow parents are slow at least partly *because* of the genes they possess. What is more, these genetic differences are transmitted to the offspring generation.

It is not inevitable that the positive slope in Figure 1.1 should have a genetic explanation. It is conceivable that fast parents are fast because they receive more nutrition than slow ones and that offspring tend to have the same dietary regime as their parents. If this were so, the positive slope would have a purely environmental explanation. Why offspring tend to resemble their parents is an empirical question. Genes are one obvious answer, but they are not the only conceivable one.

Darwin had the idea that traits are biologically inherited. However, his theory about the mechanism of inheritance—his theory of *pangenesis*—was one of many failed nineteenth-century attempts to describe the mechanism of heredity. Fortunately, Darwin's thinking about natural selection did not require that he have the right mechanism in mind; he only needed the assumption that offspring resemble their parents. Contemporary understanding of the mechanism of heredity stems from the work of Gregor Mendel (1822–1884).

I mentioned before that Darwin was not the first biologist to think that current species were descended from ancestors different from themselves. The same point can be made about the second ingredient in Darwin's theory: The idea that natural selection can modify the composition of a population was not original with Darwin. But if the idea of evolution wasn't new and the idea of natural selection wasn't new, what *was* new in Darwin's theory? Darwin's innovation was to combine these ideas—to propose that natural selection is the principal explanation of why evolution has produced the diversity of life forms we observe.

The tree of life is the pattern that evolution has produced. Natural selection, Darwin hypothesized, is the main process that explains what occurs in that tree. As mentioned before, the tree contains two kinds of events. Let us consider them in turn.

First, there is microevolution—the changes in characteristics that take place within a species. It is clear how the idea of natural selection applies to events of this sort. In the zebra example, the process begins with a population in which everyone is slow. Then, by chance, a novel organism appears. This creates the variation for natural selection to act upon. The end result is a population of fast zebras. Natural selection is only half of this process, of course. Initially, there must be variation; only then can natural selection do its work.

In this example, the change wrought by natural selection occurs within a single persisting species. A population of zebras starts the story, and the same population of zebras is around at the end. But the tree of life contains a second sort of event. Besides microevolutionary changes, there is macroevolution—new species are supposed to come into existence. How can the idea of natural selection help explain this kind of event?

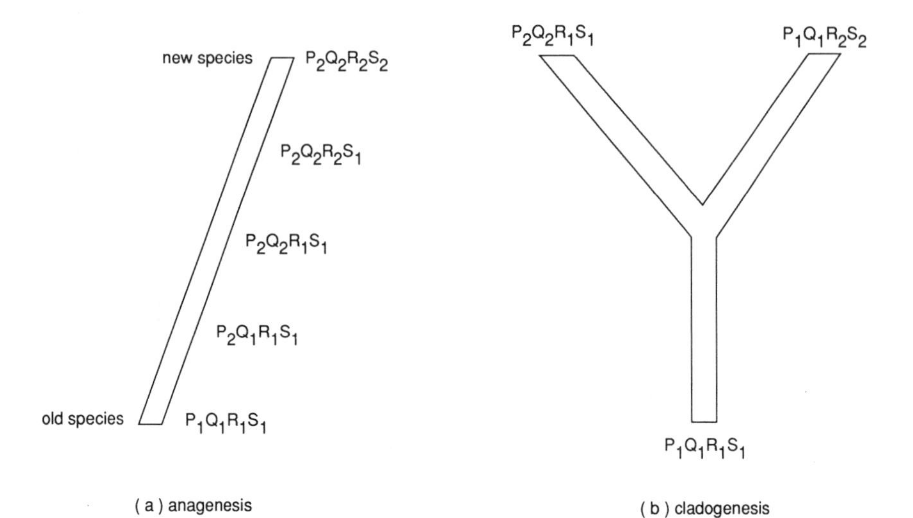

Figure 1.2 (a) In anagenesis, a single persisting lineage undergoes a gradual modification in its characteristics. (b) In cladogenesis, a parent lineage splits into two (or more) daughter lineages.

Two sorts of processes need to be considered. First, there is the idea that small changes within a species add up. A species can be made over by the gradual accumulation of evolutionary novelties. Darwin suggested that when enough such changes accumulate, ancestors and descendants should be viewed as members of different species. Notice that this process occurs within a single lineage. Modern evolutionists call it *anagenesis;* it is illustrated in Figure 1.2.

Could all speciation occur anagenetically? Not if the tree of life idea is correct. Anagenesis cannot increase the number of species that exist. An old species can go extinct, and an old species can be replaced because it gives rise to a new one. But where there was one species before, there cannot be more than one after.

Darwin envisioned a process by which species increase in numbers. This is the process of *cladogenesis,* also depicted in Figure 1.2. "Clade" is Greek for *branch;* a branching process is an indispensable part of the Darwinian picture.

How can natural selection play a role in cladogenesis? The proliferation of finches in the Galapagos Islands is a convenient example. Initially, some individuals in a species from the South American mainland were blown over to one of the islands. Further dispersal scattered these ancestors to the other islands in the Galapagos group. Local conditions varied from island to island, so natural selection led the different populations to diverge from each other. Natural selection adapts organisms to the conditions in which they exist, so when similar organisms live in different environments, the expectation is that they will diverge from each other. This is how natural selection can play an important role in the origin of species. (The two processes depicted in Figure 1.2 raise important questions about what a species is, which will be addressed in Chapter 6.)

Darwin's mechanism—natural selection—is most obviously at work in the small-scale changes that take place in a single lineage. However, Darwin conjectured that natural selection did far more than make modest modifications in the traits of existing species. He thought it was the key to explaining the origin of species. Yet, Darwin never observed a speciation event take place; nor did he observe natural selection produce a new species. If he did not observe such events, how could he possibly claim to have discovered that species evolve by the process he had in mind?

One line of argument involved the fact that plant and animal breeders had been able to modify the characteristics of organisms by artificial selection. Darwin reasoned that if breeders could change domesticated organisms so profoundly in the comparatively short span of human history, then natural selection would be able to produce far more profound changes in the longer reaches of geological time. His theory was based, in part, on the idea that if a process can produce small changes in a short period of time, it will be able to produce large changes given longer lengths of time.

Earlier, I explained that modern evolutionary theory draws a distinction between microevolution and macroevolution. The former includes the modification of traits within existing species; the latter covers the origin and extinction of species. Darwin thought that a single mechanism was fundamental to both micro- and macro-level processes. This was a bold extrapolation from the small to the large. Such extrapolations can sometimes lead to falsehood. It is not inevitable that events on different time scales must have the same explanation.

In proposing this extrapolation, Darwin was going against an influential biological idea—that there are limits beyond which a species cannot be pushed. It is easy to tinker with relatively minor features of a species. For example, this is how artificial selection produced the different dog varieties. But could selection operating on the members of a species produce a new species? Darwin went against the idea that species are fixed when he answered yes.

It is important to see that this disagreement about the malleability of species was not settled by any simple observation in Darwin's lifetime. This does not mean that Darwin had no evidence for his position; it means that his argument was more complex than might first appear.

Matters are much more straightforward now. Modern biologists have observed speciation events. Indeed, they have even caused them. As will be discussed in Chapter 6, one standard (though not uncontroversial) idea about species is that they are reproductively isolated from each other. Two contemporary populations are said to belong to different species if they cannot produce viable fertile offspring with each other. Botanists have found that the chemical *colchicine* causes *ploidy*—i.e., a modification in the number of chromosomes found in an organism. For example, by administering colchicine, a botanist can produce tetraploid plants that are reproductively isolated from their diploid parents. The daughter and parent populations satisfy the requirement of reproductive isolation. We now have *observational* evidence that species boundaries are not cast in stone.

In summary, Darwin advanced a claim about pattern and a claim about process. The pattern claim was that all terrestrial organisms are related genealogically; life forms a tree in which all contemporary species have a common ancestor if we only go back far enough in time. The process claim was that natural selection was the principal cause of the diversity we observe among life forms. However, neither of these claims was the straightforward report of what Darwin *saw*. This raises the question of how a scientist can muster evidence for hypotheses that go beyond what is observed directly. I'll address this problem in Chapter 2.

1.4 Historical Particulars and General Laws

Some sciences try to discover general laws; others aim to uncover particular sequences of historical events. It isn't that the "hard" sciences only do the former and the "soft" sciences strive solely for the latter. Each broad discipline contains subareas that differ in how they emphasize one task or the other.

Within physics, compare the different research problems that a particle physicist and an astronomer might investigate. The particle physicist might seek to identify general principles that govern a certain sort of particle collision. The laws to be stated describe what the outcome of such a collision would be, no matter *where* and no matter *when* it takes place. It is characteristic of our conception of laws that they should be *universal;* they are not limited to particular regions of space and time.

Laws take the form of if/then statements. Isaac Newton's universal Law of Gravitation says that the gravitational attraction between any two objects is directly proportional to the product of their masses and inversely proportional to the square of the distance between them. The law does not say that the universe contains two, four, or any number of objects. It just says what would be true *if* the universe contained objects with mass.

In contrast, astronomers typically will be interested in obtaining information about a unique object. Focusing on a distant star, they might attempt to infer its temperature, density, and size. Statements that provide information of this sort are not if/then in form. Such statements describe historical particulars and do not state laws.

This division between *nomothetic* ("nomos" is Greek for *law*) and *historical* sciences does not mean that each science is exclusively one or the other. The particle physicist might find that the collisions of interest often occur on the surface of the sun; if so, a detailed study of that particular object might help to infer the general law. Symmetrically, the astronomer interested in obtaining an accurate description of the star might use various laws to help make the inference.

Although the particle physicist and the astronomer may attend to both general laws and historical particulars, we can separate their two enterprises by distinguishing *means* from *ends*. The astronomer's problem is a historical one because the goal is to infer the properties of a particular object; the astronomer uses laws only as a means. Particle physics, on the other hand, is a nomothetic discipline because the goal is to infer general laws; descriptions of particular objects are relevant only as a means.

The same division exists within evolutionary biology. When a systematist infers that human beings are more closely related to chimps than they are to gorillas, this phylogenetic proposition describes a family tree that connects three species. The proposition is logically of the same type as the proposition that says that Alice is more closely related to Betty than she is to Carl. Of course, the family tree pertaining to species connects *bigger* objects than the family tree that connects individual organisms. But this difference merely concerns the size of the objects in the tree, not the basic type of proposition that is involved. Reconstructing genealogical relationships is the goal of a *historical science.*

The same can be said of much of paleobiology. Examining fossils allows the biologist to infer that various mass extinctions have taken place. Paleobiologists identify which species lived through these events and which did not. They try to explain why the mass extinctions took place. Why did some species survive while others did not? In similar fashion, a historian of our own species might try to explain the mass death of South American Indians following the Spanish Conquest. Once again, the units described differ in size. The paleobiologist focuses on whole species; a historian of the human past describes individual human beings and local populations.

Phylogenetic reconstruction and paleobiology concern the distant past. But historical sciences, as I am using that term, often aim to characterize objects that exist in the present as well. A field naturalist may track gene or phenotypic frequencies in a particular population. This is what Kettlewell (1973) did in his investigation of industrial melanism in the peppered moth (*Biston betularia*). The project was to describe and explain a set of changes. Field naturalists usually wish to characterize particular objects, not to infer general laws.

Are there general laws in evolutionary biology? Although some philosophers (Smart 1963; Beatty 1981) have said no, I want to point out that there are many interesting if/then generalizations afoot in evolutionary theory.

Biologists usually don't call them "laws"; "model" is the preferred term. When biologists specify a model of a given kind of process, they describe the rules by which a system of a given kind changes. Models have the characteristic if/then format that we associate with scientific laws. These mathematical formalisms say what will happen *if* a certain set of conditions is satisfied by a system. They do not say *when* or *where* or *how often* those conditions are satisfied in nature.

Consider an example. R. A. Fisher (1930), one of the founders of population genetics, described a set of assumptions that entails that the sex ratio in a population should evolve to 1:1 and stay there. Mating must be at random, and parental pairs must differ in the mix of sons and daughters they produce (and this difference must be heritable). Fisher was able to show, given his assumptions, that selection will favor parental pairs that produce just the minority sex. For example, if the offspring generation has more males than females, a parental pair does best by producing all daughters. If the population sex ratio is biased in one direction, selection favors traits that reduce that bias. The result is an even mix of males and females.

Fisher's model considers three generations—parents produce offspring who then produce grandoffspring. What mix of sons and daughters should a parent produce if she is to maximize the number of grandoffspring she has? If there are N individuals in the grandoffspring generation and if the offspring generation contains m males and f females, then the average son has N/m offspring and the average daughter has N/f offspring. So individuals in the offspring generation who are in the minority sex on average have more offspring. Hence, the best strategy for a mother is to produce offspring solely of the minority sex. On the other hand, if the sex ratio in the offspring generation is 1:1, a mother cannot do better than the other mothers in the population by having an uneven mix of sons and daughters. A 1:1 sex ratio is a stable equilibrium. A more exact description of Fisher's argument is provided in Box 1.3.

Fisher's elegant model is mathematically correct. If there is life in distant galaxies that satisfies his starting assumptions, then a 1:1 sex ratio must evolve. Like Newton's universal Law of Gravitation, Fisher's model is not limited in its application to any particular place or time. And just as Fisher's model may have millions of applications, it also may have none at all. The model is an if/then statement; it leaves open the possibility that the *if*s are never satisfied. Field naturalists have the job of saying whether Fisher's assumptions apply to this or that specific population.

In deciding whether something is a law or a historical hypothesis, one must be clear about which *proposition* one wishes to classify. For example, to ask whether "natural selection" is a law is meaningless until one specifies which proposition about natural selection is at issue. To say that natural selection is responsible for the fact that human beings have opposable thumbs is to state a historical hypothesis; but to say that natural selection will lead to an even sex ratio in the circumstances that Fisher described is to state a law. (Evolutionary laws will be discussed further in Section 3.4.)

Although inferring laws and reconstructing history are distinct scientific goals, they often are fruitfully pursued together. Theoreticians hope their models are not vacuous; they want them to apply to the real world of living organisms. Likewise, naturalists who describe the present and past of particular species often do so with an eye to providing data that have a wider theoretical significance. Nomothetic and historical disciplines in evolutionary biology have much to learn from each other.

An example of a particularly recalcitrant problem in current theory may help make this clear. We presently do not understand why sexual reproduction is as prevalent as it is. The problem is not that theoreticians cannot write models in which sexual reproduction is advantageous. There are lots of such models, each of them mathematically correct. Indeed, there also are many models that show that under specified conditions, sex will be *dis*advantageous.

Box 1.3 Fisher's Sex Ratio Argument

The accompanying text provides a simplified rendition of Fisher's argument. In point of fact, Fisher did not conclude that there should be *equal numbers* of males and females but that there should be *equal investment*. A mother has a total package of energy (T) that she can use to produce her mix of sons and daughters. Suppose p is the percentage of energy she allots to sons, that each son costs c_m units of energy to raise, and that a son brings in b_m units of benefit. With a similar representation of the costs and benefits of daughters, a mother's total benefit from her sons and daughters is

$$b_m[pT/c_m] + b_f[(1-p)\,T/c_f].$$

Suppose all mothers in the population allocate p and $(1-p)$ of their resources to sons and daughters, respectively. When will a mutant mother do better by departing from this behavior—i.e., by allotting p^* and $(1-p^*)$ to sons and daughters (where $p \neq p^*$)? This novel mother does better than the residents precisely when

$$b_m[p^*T/c_m] + b_f[(1-p^*)\,T/c_f] > b_m[pT/c_m] + b_f[(1-p)\,T/c_f],$$

which simplifies to

$$(b_m/c_m - b_f/c_f)\,(p^* - p) > 0.$$

Recall from the accompanying text that a son provides a benefit of N/m and a daughter provides a benefit of N/f. Substituting these for the benefit terms in the above expression, we obtain

$$(N/mc_m - N/fc_f)(p^* - p) > 0.$$

When the residents invest equally in sons and daughters ($mc_m = fc_f$), no mutant strategy can do better than the resident strategy. And when the residents invest *un*equally, a mutant will do better than the residents by investing exclusively in the sex in which the residents have *under*invested.

How does investment in the two sexes affect the numbers of sons and daughters produced? In human beings, males have a higher mortality rate, both prenatally and postnatally. This means that the average son costs less than the average daughter. In this case, equal investment entails that an excess of males is produced at birth, which is what we observe.

Fisher's argument assumes that there is random mating in the offspring generation. The import of this assumption was first explored by Hamilton (1967). If there is strict brother/sister mating, then a parent maximizes the number of grandoffspring she has by producing a female-biased sex ratio among her progeny.

The difficulty is not that the models are wrong as if/then statements but that they often fail to apply to nature. In the real world, some species are sexual, whereas others are not. These different species live under a variety of conditions, and their phylogenetic backgrounds differ as well. What we would like is a model that fits the diversity we observe. To date, no model can claim to do this.

If model building (the pursuit of laws) proceeded independently of natural history, the evolution of sex would not be puzzling. A model can easily show how sex *might* have evolved; *if* the assumptions of the model were satisfied by some natural population, that population would evolve a sexual mode of reproduction. It is a historical question whether this or that population actually satisfied the assumptions in the model. Only by combining laws and history can one say why sex *did* evolve.

1.5 The Causes of Evolution

Although the data of natural history are indispensable to evolutionary model building, there is a place for model building that floats free from the details of what we have observed. Fisher (1930, pp. VIII–IX) put the point well when he remarked that "no practical biologist interested in sexual reproduction would be led to work out the detailed consequences experienced by organisms having three or more sexes; yet what else should he do if he wishes to understand why the sexes are, in fact, always two?" We often understand the actual world by locating it in a broader space of possibilities.

Models map out the possible causes of evolution. What are these possible causes? I have already mentioned *natural selection;* heritable variation in fitness can produce evolution. And in Section 1.1, I explained how the *system of mating* in a population can modify the frequencies of different genotypes. There are other possible causes as well.

Gene frequencies can change because of *mutation*. A population that is 100 percent *A* can evolve away from this homogeneous state if *A* genes mutate into *a*. A model of the mutation process considers both the rate of forward mutation (from *A* to *a*) and the rate of backward mutation (from *a* to *A*). When the only influences on the gene frequencies at this locus are these two mutation rates, the population will evolve to an equilibrium gene frequency that is determined just by the mutation rates.

Another possible cause of evolution is *migration*. Migrants may move into and out of a population. The situation is similar to the one described in models of mutation pressure. The rates of genes flowing in and of genes flowing out can move the population gene frequency to an equilibrium value.

Random genetic drift also can modify gene frequencies (Kimura 1983). Consider a haploid population in which there are 100 individuals, each at the juvenile stage; at a given locus, 50 percent have the *A* gene and 50 percent have *a*. Suppose that these individuals have the same chance of surviving to adulthood. Does this mean that the gene frequency at adulthood must be precisely 50/50? The answer is no. To say that *A* individuals have the same chance of surviving as *a* individuals does not mean that they must do so in exactly equal

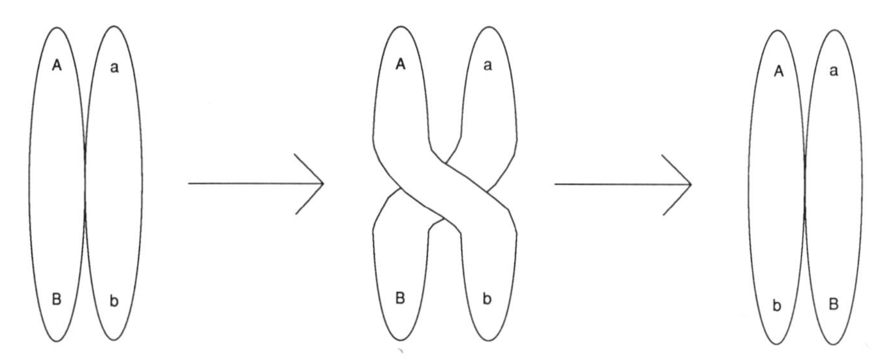

Figure 1.3 A double heterozygote undergoing recombination by crossing over.

numbers. If a fair coin is tossed, heads has the same chance of landing face up as tails does. But that does not mean that in a run of 100 tosses, there must be exactly 50 heads and 50 tails. By the same token, genes in a population may be selectively equivalent and still change their frequencies because of *chance*.

I mentioned in Section 1.1 that the definition of evolution as change in gene frequency is too restrictive. Evolution also can occur when there is change in the frequencies of various combinations of genes. *Recombination* is an important process that can cause this to happen. Consider a diploid individual that is heterozygous at both the *A*- and the *B*-locus. As depicted in Figure 1.3, this individual has *A* and *B* on one chromosome and *a* and *b* on the other. Recombination occurs when the two chromosomes *cross over*. The result is that *A* and *b* end up on the same chromosome, as do *a* and *B*.

Suppose a population begins with every individual homozygous for *A* at the *A*-locus and homozygous for *B* at the *B*-locus. Then there is a mutation at the *A* locus; a copy of *a* makes its appearance. Following that, in another organism, there is a mutation at the *B*-locus, which introduces a copy of *b*. The three organismic configurations now present in the population are shown in Figure 1.4.

These three types of individuals subsequently breed with one another, each forming gametes to do so. Notice that there are only *three* gametic types in the population; a gamete can be *AB*, *aB*, or *Ab*. Without recombination, no gamete will be *ab* (which means that *a* and *b* can't both go to 100 percent representation in the population). Recombination is an important process because it can enrich the range of variation. Mutation produces new *single* genes; recombination produces new *combinations* of genes on the same chromosome.

The causes just listed need not occur alone. Within a given population, selection, mutation, migration, recombination, pattern of mating, and drift all may simultaneously contribute to the changes in frequency that result. Simple models of evolution describe what will happen when one of these forces acts alone. More complicated models describe how two or more of these forces act simultaneously. Since populations in the real world are impinged upon by a

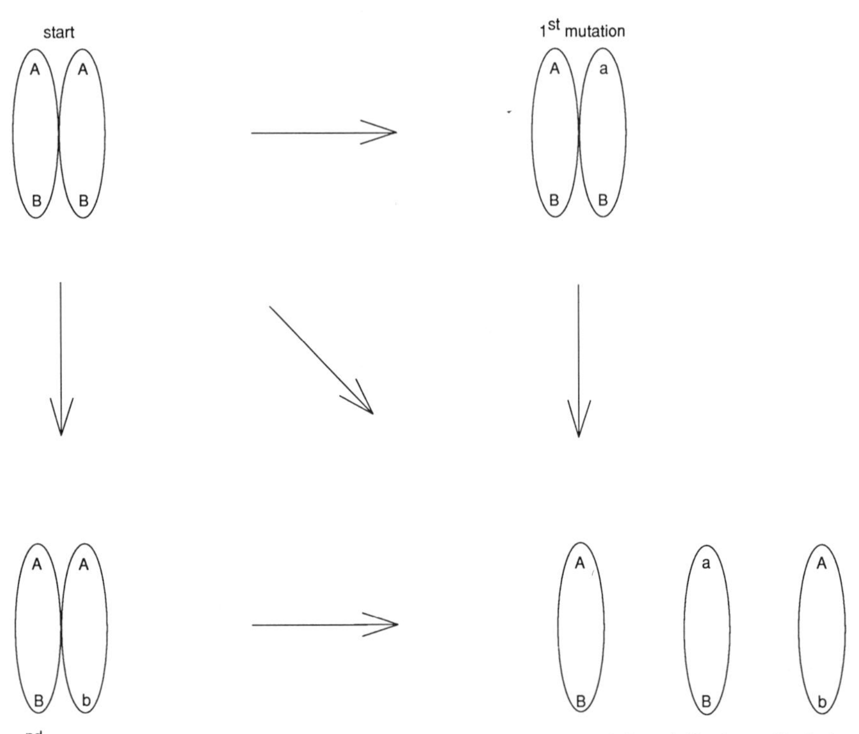

Figure 1.4 Gamete formation without recombination. Note that no gamete contains both a and b.

multiplicity of causes, complicating a model by taking account of more variables is a way to make the model more realistic.

Besides identifying the consequences that these causes of evolution may have for the composition of a population, evolutionary biology also describes what can bring these causes into being. Mutations can cause evolution, but what causes mutation? We currently know a good deal about the way mutagens in the environment (radiation from the sun, for example) can produce mutations. In addition to understanding the *consequences* of mutation, we also have some understanding of its *sources*.

The same double-aspect understanding is available for other causes of evolution. A population geneticist can describe what will happen to the gene frequencies at a locus when individuals with different genotypes vary with respect to their abilities to survive and reproduce. Models of this sort describe the consequences of fitness differences. A separate question concerns the sources of selection: When will natural selection favor one variant over another?

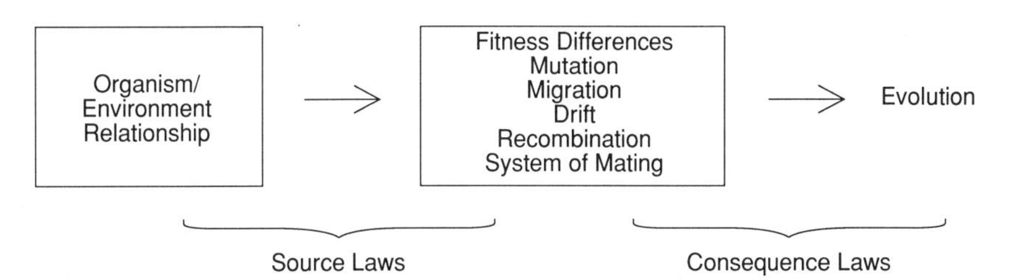

Figure 1.5 Models in evolutionary biology provide both source laws and consequence laws for the causes of evolution.

I have already cited Fisher's model of sex ratio evolution, which describes how the mix of sons and daughters produced by a parental pair affects the pair's reproductive success. This model describes how phenotypic differences among organisms can generate differences in fitness. Another example concerns the contrast between organisms that are *specialists* and organisms that are *generalists*. Generalists make a living in a number of ways; specialists are more limited in what they do, although they often are better within their specialty than a generalist is. Intuitively, an organism that lives in a heterogeneous environment will do better as a generalist, and one that lives in a homogeneous environment will be better off specializing. Here, we are describing how relationships between organism and environment can lead natural selection to favor some variants over others. Ideas such as this one describe the sources of selective differences.

In summary, models in evolutionary theory describe both the sources and the consequences of the different causes of evolution. This division of labor between the two theoretical undertakings is shown in Figure 1.5.

Population geneticists often work out the consequence laws of evolution. Once the magnitudes of the various causes are specified, a population genetics model allows one to compute the evolutionary consequences. It is not part of such models to say *why* one genotype is fitter than another or *why* there is a difference between the forward and backward mutation rates at some locus. Evolutionary ecology, on the other hand, often aims to formulate evolutionary models concerning the sources of evolutionary pressures.

As noted in the previous section, the main reason to construct evolutionary models about the possible causes of evolution is to apply them to the actual world. We wish to know not just what *can* cause evolution but what has, in fact, done so. We can pose this as a question about a single trait in a single species ("Why do polar bears have white fur?"). We also can pose it as a question about several species, inquiring as to why these species differ from each other ("Why do Indian rhinoceri have one horn and African rhinoceri have two?").

One of the most controversial matters in current evolutionary theory concerns the importance of natural selection as a cause of evolution. It is obvious that selection is a *possible* cause; the question is, How important has it been in the *actual* course of evolution? Many evolutionary biologists automatically look for explanations in terms of natural selection; others think that the importance of natural selection has been exaggerated and that the reasoning that backs many selectionist explanations has been sloppy. This debate about *adaptationism* will be discussed in Chapter 5.

1.6 The Domains of Biology and Physics

Physics is about any and all objects that are *made of matter*. Biology is about objects that are *alive*. And psychology is about objects that have *minds*. Although all of these claims require some fine-tuning, each is roughly accurate. Each describes the *domain* of the science in question. Each tells you what class of objects you should consider if you want to decide whether a proposed generalization in physics, biology, or psychology is correct.

How are these domains related to each other? Let's begin with the relationship of biology and physics. Figure 1.6(a) depicts two proposals. The first, which I will call *physicalism*, claims that all living things are physical objects. If you take an organism, no matter how complex, and break it down into its constituents, you will find matter and only matter there. Living things are made of the same basic ingredients as nonliving things. The difference is in how those basic ingredients are put together.

Vitalism, at least in some of its formulations, rejects this physicalistic picture. It says that living things are alive because they contain an immaterial ingredient—an *élan vital* (Henri Bergson's term) or an *entelechy* (the Aristotelian term used by Hans Driesch). Vitalism therefore maintains that some objects in the world are not purely physical.

According to vitalism, two objects could be physically identical even though one of them is alive while the other is not. The first could contain the life-giving immaterial ingredient while the second fails to do so. Physicalists scoff at this. They maintain that if two objects are physically identical, they must have all the same biological properties; either both are alive or neither is (a point to which I will return in Section 3.5).

Vitalism is easiest to take seriously when science is ignorant of what lies behind various biological processes. For example, before the physical basis of respiration was understood, it was possible to suggest that organisms are able to breathe only because they are animated by an immaterial life principle. Similarly, before molecular biology explained so much about the physical basis of heredity, it was possible to entertain vitalistic theories about how parents influence the characteristics of their offspring. The progress of science has made such claims about respiration and inheritance wildly implausible.

Still, there are problems in biology that remain unsolved. The area of development (ontogeny) is full of unanswered questions. How can a single-celled embryo produce an organism in which there are different specialized cell types? How do these cell types organize themselves into organ systems? No adequate physicalistic explanation is available now, so why not advance a vitalistic claim about ontogenetic processes? The point to recognize is that vitalism does not become plausible just because we currently lack a physical ex-

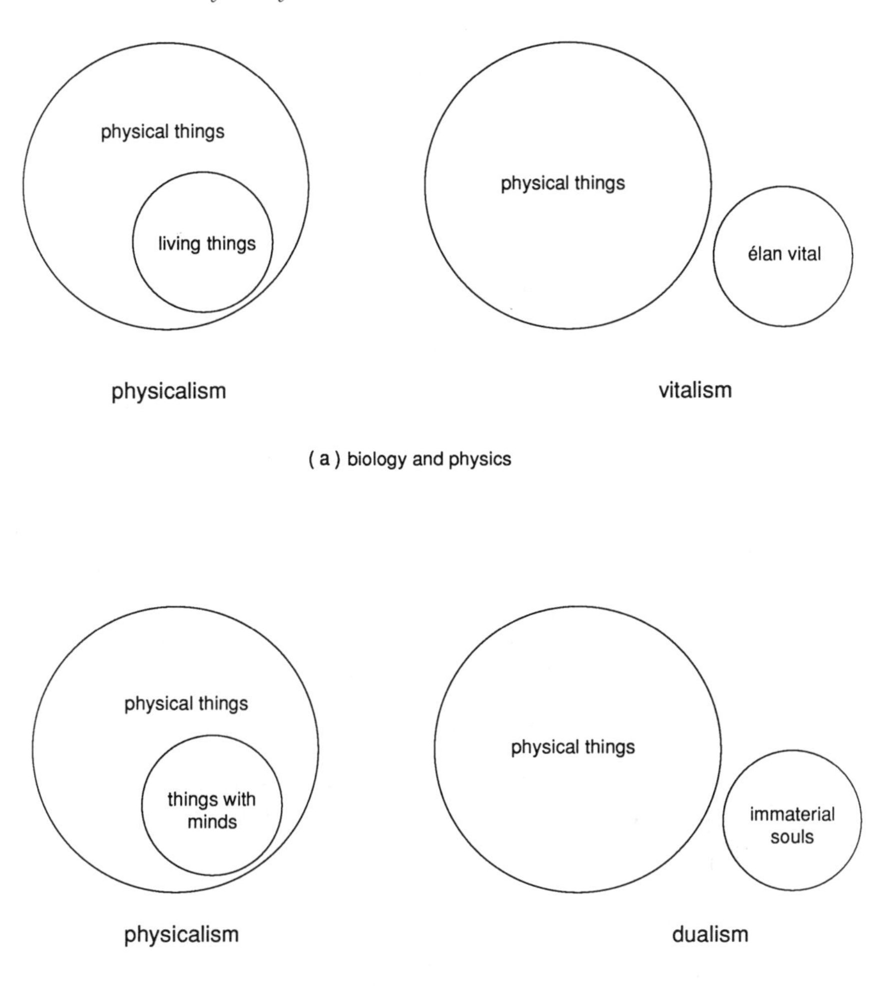

Figure 1.6 (a) Physicalism maintains that all living things are made of matter and of nothing else, whereas vitalism asserts that living things contain an immaterial substance—an *élan vital.* (b) In the philosophy of mind, physicalists and dualists disagree about whether the mind is made of an immaterial substance.

planation. If vitalism is to be made plausible, a more direct line of defense must be provided.

Another special feature of living things is worth considering. Organisms are goal-directed (teleological) systems; they act so as to further their ends of surviving and reproducing. Does this observation require us to posit the existence of an immaterial ingredient in living things that directs them toward what they need? As we will see in Section 3.7, the theory of natural selection allows us to formulate an explanation of this fact about organisms that does not require vitalism.

Vitalism is held in low repute by biologists today because no strong positive argument on its behalf has ever been constructed. In addition, the progress of science has enormously increased our understanding of the physical bases of life processes. It is a sound working hypothesis (which may just possibly turn out to be mistaken) that living things are nothing but structured chunks of matter.

There is an interesting parallelism between the issue of vitalism and the issue in the philosophy of mind called the mind/body problem (Figure 1.6[b]). How is the domain of psychology, which includes any object that has a mind, related to the domain of physics? Physicalism maintains that each and every object that possesses psychological properties is a physical thing. Mind/body dualism, in contrast, maintains that the mind is an immaterial substance, distinct from the body.

In the seventeenth century, René Descartes produced a few ingenious arguments in favor of dualism. This is not the place to review them, but I will note that Descartes did not rest his case on the fact that the physics of his time could not explain the mind in all its aspects. He tried to provide a positive argument that the mind and the body are distinct.

The main difficulty for dualism has been to account for the apparent causal interactions that exist between the mental and the physical. For instance, taking aspirin makes headaches go away, and people's beliefs and desires can send their bodies into motion. If the mind is immaterial, then it does not take up space. But if it lacks spatial location, how can it be causally connected to the body? When two events are causally connected, we normally expect there to be a physical signal that passes from one to the other. How can a physical signal emerge from or lead to the mind if the mind is no place at all?

Because of difficulties of this sort, dualists have sometimes abandoned the idea that mind and body causally interact. They try to argue that taking aspirin is followed by the diminishment of headaches even though no causal process links the two. However, this fallback position faces a difficulty of its own. Without causal connections, the many regularities that link mind and body seem to be cosmic coincidences. Surely it is preferable to be able to explain such regularities as the result of causal connections. Physicalism is able to make sense of these causal connections; dualism has never been able to do so. Just as this point favors physicalism in the mind/body problem, it also supports physicalism over vitalism in biology.

1.7 Biological Explanations and Physical Explanations

Adopting a physicalistic view of the domain of biology simply means that one accepts the idea that living things are physical objects. It is important to realize that this thesis does not say what the relationship is between *biological explanations* and *explanations in physics*. Even if living things are made of matter and nothing else, the fact remains that the vocabulary of biology radically differs from that of physics. Physicists talk about elementary particles, space-time, and quantum mechanical states; evolutionary biologists talk about phylogenies, ecosystems, and inbreeding coefficients. Even though the domain of biology falls within the domain of physics, the vocabulary of biology and the vocabulary of physics have little overlap. Explanations in biology are produced in the distinctive vocabulary of biology; explanations in physics use the distinctive vocabulary of physics. The question is how these two kinds of explanation fit together.

It is quite clear that physics explains some facts that do not have a biological explanation. For example, biology has nothing significant to contribute to our understanding of why planets move in elliptical orbits. But now let us consider the relation of biology to physics from the other direction: Is it true that every fact explained by biology also can be explained by physics? This is one way to ask whether biology *reduces* to physics.

If it is said that everything in biology *can* be explained by physics, does this mean *can in principle* or *can in practice*? Explainability-in-principle means that an ideally complete physics would be able to account for all biological phenomena. Explainability-in-practice means that we can explain all biological phenomena with the physics we currently possess.

How might current physics be applied to problems in biology? Clearly, there are many areas of biology for which we have no clue how to do this. I've already mentioned two—the evolution of sex and the field of ontogenetic development. Although there is no reason to doubt that these phenomena *are consistent with* our current best physical theories, no one has the slightest idea how the physics might be put to work.

Even with respect to phenomena that are well understood biologically, scientists have rarely bothered to work out how physics might be used to provide explanations. Fisher's sex ratio argument again furnishes a useful example. Although this model allows us to understand why the sex ratio is 1:1 in some populations, it says absolutely nothing about the nature of the matter out of which organisms are composed. Again, there is no reason to deny that Fisher's model *is consistent with* currently accepted physics; but it is entirely unclear how modern physics could be used to explain what Fisher's model allows us to explain.

Let us shift the question, then, to the realm of reducibility-in-principle. Once we understand the evolution of sex (or some other phenomenon) within the framework of evolutionary theory (assuming that this will happen!), once we have a fully adequate set of physical theories, and once we see how these physical and biological theories connect with each other, will we then be able to explain the evolution of sex from the point of view of physics?

There are several *if*s in this question. And notice that even if the answer to the question is yes, that would say nothing much about how current research should be conducted. Even if biology is *in principle* reducible to physics, this does not mean that the best way to advance our present understanding of biological problems is to think about quarks and space-time. Perhaps a completed science would be able to unite biology and physics, but this claim about some hypothetical future says nothing about how we should conduct our investigations in the present.

So the thesis of reducibility-in-principle does not seem to have many direct methodological consequences for current scientific practice. Still, it is of some philosophical interest because we would like to understand how the goals of the different sciences mesh together. I'll discuss this problem further in Section 3.5.

Suggestions for Further Reading

Maynard Smith (1977) and Futuyma (1986) are good introductions to evolutionary biology. Ruse (1973), Hull (1974), and Rosenberg (1985) are worthwhile introductions to the philosophy of biology, and Ruse (1988b) provides an excellent bibliographic survey of recent work in this field. Williams (1973) develops an axiomatic approach to the structure of evolutionary theory, discussed in Sober (1984a). Beatty (1981, 1987), Lloyd (1988), and Thompson (1988) argue that the semantic view of scientific theories helps illuminate the structure of evolutionary theory; Ereshefsky (1991a) criticizes the semantic view.

2

Creationism

2.1 The Danger of Anachronism

To understand the history of an idea, we must avoid reading our present understanding back into the past. It is a mistake to assume that an idea we now regard as unacceptable was never part of genuine science in the first place.

Consider, for example, the claim that phrenology is a pseudoscience. Although I would doubt the seriousness of someone who believes in phrenology today, the fact remains that it was a serious research program in the nineteenth century. The program was guided by three main tenets. First, phrenologists held that specific psychological characteristics are localized in specific regions of the brain. Second, they held that the more of a given talent or psychological tendency you possess, the bigger that part of your brain will be. Third, they held that the bumps and valleys on the skull reflect the contours of the brain. Given these three ideas, they reasoned, it should be possible to discover people's mental characteristics by measuring the shape of their skulls.

Phrenologists disagreed among themselves about which mental characteristics should be regarded as fundamental and about where those characteristics were localized. Is fear of snakes a trait, or is the more general characteristic of fearfulness the one that has neurological reality? If fearfulness is the right trait to consider, to what detail of brain shape does it correspond? Phrenologists made little progress on these problems. Various versions of phrenology were developed, but each failed to receive serious empirical confirmation. After a while, the research program ground to a halt. It eventually became reasonable to discard the program because the field had failed to progress.

Contemporary brain scientists looking back at phrenology might be tempted to see skull measuring as a pseudoscience. The point I wish to emphasize is that what is true *now* was not true *then:* Today, we have serious evidence against at least the second and third tenets of the phrenological research program. But this does not mean that individuals working in that framework were not doing science. Their ideas were false, but it is anachronistic to expect them to have known what we know now.

Suppose a group of people *now* were to defend phrenological ideas, ignoring the wealth of evidence we currently have against phrenological theories and insisting dogmatically that bumps on the head really do reveal what a person's mind is like. Would this group's ideas count as pseudoscience? Here, we must be careful. We must distinguish the *people* from the *propositions* they maintain. These present-day phrenologists are pigheaded. They behave in a way that could be called unscientific. But this does not mean that the propositions they defend are not scientific. These people are endorsing a theory that has been refuted by ample scientific evidence. The propositions are scientific in the sense that they are scientifically testable.

I just said that people behave unscientifically when they refuse to consider relevant evidence. This does not mean that scientists never behave pigheadedly. They are people too; they can exemplify all of the failings to which nonscientists are subject. Scientists *ought* to avoid being pigheaded; it is another question how often they succeed in doing this.

It is one thing to say that a person is behaving unscientifically, quite another to say that the theory the person defends is not a scientific proposition at all. Indeed, a person can behave pigheadedly toward propositions that are perfectly scientific. For instance, the proposition that *the earth is flat* is a scientific proposition. It can be tested by scientific means, which is why we are entitled to regard it as false. Yet, flat-earthers are not behaving scientifically when they dogmatically accept this perfectly testable proposition even though there is lots of evidence against it.

These remarks about phrenologists and flat-earthers are intended to set the stage for some of the principal conclusions I will reach about creationism. Creationists maintain that species were separately created by God. Species are not genealogically related; they did not evolve from common ancestors under the influence of natural selection. Is creationism a scientific theory? If so, why do scientists fail to take it seriously? Creationists claim that scientists fail to be open-minded when they dismiss the hypothesis of intelligent design. Are evolutionary biologists therefore guilty of unscientific pigheadedness?

Creationists press these questions because they have a political agenda. They wish to reduce or eliminate the teaching of evolution in high school biology courses and to have the biblical story of creation taught in the public schools. As a strategic matter, they realize that they cannot admit that their views are religious in nature. To do so would frustrate their ambitions since the U.S. Constitution endorses and the courts have supported a principled separation of church and state. To avoid this problem, they have invented the term "scientific creationism." Scientific creationists attempt to defend creationism by appeal to evidence, not by appeal to biblical authority. If theirs is a scientific theory that is just as well supported as evolutionism, then creationists can argue that the two theories deserve "equal time."

The preceding paragraph attempts to describe the motives that creationists have. However, I do not want to concentrate on the motives of creationists *or* of evolutionists. I am interested in assessing the logic of the positions they defend, not their motives for defending them. My focus will be on *propositions,* not *people.*

I believe that some of the hypotheses of creationism are testable. This theory is like the doctrines of phrenology and the ideas of flat-earthers. If this is right, then the reason for keeping creationism out of the public schools is not that creationist theories are Religion (with a capital *R*), while biology courses are devoted to Science (with a capital *S*). Rather, creationism is similar to other discredited theories that do not deserve a central place in biology teaching. We exclude the ideas of phrenologists and flat-earthers not because the ideas are unscientific but because they have been refuted scientifically. Equal time is more than creationism deserves.

All the same, I do not think that creationism should be airbrushed from the history of evolutionary thought; it is not a subject that should go unmentioned in science education. To grasp the power of evolutionary thinking, it is important to understand how the theory of evolution emerged historically. Creationism was an influential idea with which the theory of evolution competed. Creationism should be taught but not because it is a plausible candidate for the truth: It should be described so that its failures are patent.

What now goes by the name of creationism is the fossilized remains of what once was a vital intellectual tradition. In my opinion, current books and articles produced by creationists are not serious contributions to thought. However, the same is not true of the writings of earlier defenders of creationism. Before Darwin's time, some of the best and the brightest in both philosophy and science argued that the adaptedness of organisms can be explained only by the hypothesis that organisms are the product of intelligent design. This line of reasoning—the *design argument*—is worth considering as an object of real intellectual beauty. It was not the fantasy of crackpots but the fruits of creative genius.

Here, I must remind the reader of the dangers of anachronistic thinking. For those who doubt the intellectual seriousness of present-day creationism, it is tempting to think that the theory was never a serious position. Evolutionists view contemporary creationists as purveyors of pseudoscience, and evolutionists sometimes conclude from this that creationism has always been opposed to a scientific worldview.

To understand the power that the design argument once had, it is essential to suspend for the moment the familiar modern idea that scientific and religious modes of thought stand in fundamental opposition to one another. It is quite common now for people who take religion seriously to insist that religious convictions are based on faith, not reason. However, this opposition is entirely alien to the tradition of rational theology, which seeks to put religious conviction on a rational footing. It was within this tradition that much of what is best in Western philosophy was written. The design argument was intended as a "scientific argument." What I mean by this quoted expression will become clear presently.

2.2 Paley's Watch and the Likelihood Principle

In the *Summa Theologiae,* St. Thomas Aquinas (1224–1274) presented five ways to prove that God exists. The fifth of these was the argument from design. Aquinas's version of the design argument elaborated ideas already put forward by Plato and Aristotle. Yet, for all its long history, the true heyday of the design argument came later. Principally in Britain and from the time of the scientific revolution to the publication of Darwin's *Origin of Species* (1859), the argument from design enjoyed a robust life. A number of talented thinkers developed it, finding new details they could embed in its overall framework.

Many philosophers now regard David Hume's *Dialogues Concerning Natural Religion* (1779) as the watershed in this argument's career. Before Hume, it was possible for serious people to be persuaded by the argument, but after the onslaught of Hume's corrosive skepticism, the argument was in shambles and remained that way forever after.

Biologists with an interest in the history of this idea often take a different view (Dawkins 1986), seeing the publication of Darwin's *Origin of Species* as the watershed event. For the first time, a plausible, nontheistic explanation of adaptation was on the table. After Darwin, there was no longer a need to invoke intelligent design to explain the adaptedness of organisms.

Creationists, of course, take a third view of this historical question. They deny that the argument died at the hands of either Hume *or* Darwin since they think it is alive and well today.

It is possible to pose the question about the history of the design argument in two ways. The first is sociological: When (if ever) did educated opinion turn against the design argument? With respect to this question, it is quite clear that Hume's *Dialogues* did not put a stop to the argument. In the years between Hume's posthumous publication and the appearance of the *Origin of Species,* the argument fostered a cottage industry. A series of volumes called the *Bridgewater Treatises* appeared, in which some of the best philosophers and scientists in Britain took the design argument very seriously indeed.

However, this sociological fact leaves unanswered the second historical question we can ask about the design argument. When (if ever) was the argument shown to be fatally flawed? Many philosophers nowadays think that Hume dealt the deathblow. In their view, the ideas presented in the *Bridgewater Treatises* were walking corpses; the design argument was propped up and paraded even though it already had entered *rigor mortis.*

I will consider some of Hume's criticisms of the design argument in the next section. For now, I want to identify the argument's logic. I'll discuss the version of the argument set forth by William Paley in his *Natural Theology* (1805).

The design argument is intended by its proponents to be an *inference to the best explanation* (an "abduction," in the terminology of C. S. Peirce). There are two fundamental facts about living things that cry out for explanation. Organisms are intricate and well adapted. Their complexity is not a jumble of uncoordinated parts; rather, when we examine the parts with the utmost care, we discern how the different parts contribute to the well-functioning of the organism as a whole.

Paley considers two possible explanations of these observations. The first is that organisms were created by an intelligent designer. God is an engineer who built organisms so that they would be well suited to the life tasks they face. The second possible explanation is that random physical forces acted on lumps of matter and turned them into living things. Paley's goal is to show that the first explanation is far more plausible than the second.

To convince us that the design hypothesis is better supported than the randomness hypothesis, Paley constructed an analogy. Suppose you were walking across a heath and found a watch. You open the back of the watch and observe that it is intricate and that its parts are connected in such a way that the watch as a whole is well suited to the task of timekeeping. How might you explain the existence and characteristics of this object?

One possibility is that the watch is the product of intelligent design; it is intricate and adapted to the task of timekeeping because a watchmaker made it that way. The other possibility is that random physical processes acting on a lump of metal produced the watch. Rain and wind and lightning impinged upon the lump of matter, turning it into a watch.

Which explanation of the existence and characteristics of the watch is more plausible? Paley says that the design hypothesis is far better supported by the watch's observable characteristics. He then says to the reader: If you agree with this assessment of the two hypotheses about the watch, you should draw a similar conclusion about the complexity and adaptedness of living things. In both cases, the design hypothesis is far more plausible than the randomness hypothesis.

I have interpreted Paley as constructing *two* arguments—one about a watch, the other about living things; he contends that the second argument is at least as strong as the first. The design argument, developed in this way, is an argument about two arguments.

Let's consider in more detail how each of the two arguments works. They have something important in common, even though their subject matters—watches and organisms—differ. Both arguments make use of the *Likelihood Principle* (Edwards 1972). Consider a statement we know to be true by observation; call this statement O. Then consider two possible explanations (H_1 and H_2) for why O is true. The Likelihood Principle reads as follows:

> O strongly favors H_1 over H_2 if and only if H_1 assigns to O a probability that is much bigger than the probability that H_2 assigns to O.

In the notation of probability theory, the principle says:

> O strongly favors H_1 over H_2 if and only if $P(O/H_1) >> P(O/H_2)$.

The expression "$P(O/H_1)$" represents the *likelihood* that the hypothesis H_1 has in the light of the observation O. Don't confuse this quantity with the *probability* that H_1 has in the light of O. Don't confuse $P(O/H_1)$ with $P(H_1/O)$. Although the expressions "it is likely" and "it is probable" are used interchangeably in ordinary talk, I follow R. A. Fisher in using the terms so that they mean quite different things.

How can $P(O/H)$ and $P(H/O)$ be different? Why all this fuss about distinguishing the likelihood of a hypothesis from its probability? Consider the following example. You and I are sitting in a cabin one night, and we hear rumbling in the attic. We wonder what could have produced the noise. I suggest that the explanation is that there are gremlins in the attic and that they are bowling. You dismiss this explanation as implausible.

Let O be the observation statement "there is rumbling in the attic." Let H be the hypothesis "there are gremlins in the attic, and they are bowling." I hope you see that $P(O/H)$ is very high but $P(H/O)$ is not high at all. *If* there actually were gremlins bowling up there, we would expect to hear noise. But the mere fact that we hear the noise does not make it very probable that there are gremlins bowling. The gremlin hypothesis has a high likelihood but a low probability, given the noises we hear.

This example, besides convincing you that the likelihood of a statement and the probability of a statement are different, also should convince you that there is more to a statement's plausibility than its likelihood. The gremlin hypothesis has a very high likelihood; in fact, it is arguable that no other explanation of the attic noise could have a higher likelihood. Yet, the gremlin hypothesis is not very plausible. This helps clarify what the Likelihood Principle does and does not purport to characterize.

This principle simply says whether the observations under consideration favor one hypothesis over another. It does not tell you to believe the one that is better supported by the piece of evidence under consideration. In fact, you may, in a given case, decline to believe *either* hypothesis, even though you admit that the observations favor one over the other. The Likelihood Principle does not pretend to tell you how much evidence suffices for belief. It simply provides a device for assessing the meaning of the evidence at hand.

Another issue that the Likelihood Principle does not address is the import of information *besides* the observations at hand. In the gremlin case, we know a great deal more about the world than what is encoded in proposition O. The gremlin hypothesis has a high likelihood, relative to O, but we regard this hypothesis as antecedently implausible. The overall plausibility of a hypothesis is a function both of its likelihood relative to *present* observations and its *antecedent* plausibility. The Likelihood Principle does not say that the more likely of two hypotheses (relative to some observation that one is considering) is the hypothesis with greater overall plausibility (relative to everything else one knows).

So the Likelihood Principle has quite modest pretensions. It does not tell you what to believe, and it does not tell you which of the competing hypotheses is, overall, more plausible. It simply tells you how to interpret the single observation at hand. If the first hypothesis tells you that O was to be expected while the second hypothesis says that it is almost a miracle that O was true, then O favors the first hypothesis over the second.

Now let's return to Paley's argument. I said before that his argument involves comparing two different arguments—the first about a watch, the second about living things. We can represent the statements involved in the watch argument as follows:

A: The watch is intricate and well suited to the task of timekeeping.
W_1: The watch is the product of intelligent design.
W_2: The watch is the product of random physical processes.

Paley claims that $P(A/W_1) >> P(A/W_2)$. He then says that the same pattern of analysis applies to the following triplet of statements:

B: Living things are intricate and well-suited to the tasks of surviving and reproducing.
L_1: Living things are the product of intelligent design.
L_2: Living things are the product of random physical processes.

Paley argues that if you agree with him about the watch, you also should agree that $P(B/L_1) >> P(B/L_2)$. Although the subject matters of the two arguments are different, their logic is the same. Both are inferences to the best explanation in which the Likelihood Principle is used to determine which hypothesis is better supported by the observations.

2.3 Hume's Critique

Hume did not think of the design argument in the way I have presented it. For him, the argument is not an inference to the best explanation; rather, it is an argument from analogy, or an inductive argument. This alternate conception of the argument makes a great deal of difference. Hume's criticisms are quite powerful if the argument has the character he attributes to it. But if the argument is, as I maintain, an inference to the best explanation, Hume's criticisms entirely lose their bite.

Although Paley wrote after Hume was dead, it is easy enough to reformulate Paley's argument so that it follows the pattern that Hume thought all design arguments obey. For Hume, this argument rested on an analogy between living things and artifacts:

Watches are the product of intelligent design.
Watches and organisms are similar.
═══════════
Organisms are the product of intelligent design.

I draw a double line between the premisses and the conclusion of this argument to indicate that the premisses are supposed to make the conclusion probable or highly plausible; the argument is not intended to be deductively valid. (Deductive validity means that the premisses, if true, would absolutely guarantee that the conclusion must be true.)

If the design argument is an argument from analogy, we must ask how strongly the premisses support the conclusion. Do they make the conclusion enormously plausible, or do they only weakly support it? Hume says that analogy arguments are stronger or weaker according to how similar the two objects are. To illustrate this point, he asks us to compare the following two analogy arguments:

In human beings, the blood circulates.
Human beings and dogs are similar.
═══════════
In dogs, the blood circulates.

In human beings, the blood circulates.
Human beings and plants are similar.
═══════════
In plants, the blood circulates.

The first argument, Hume says, is far stronger than the second because human beings are much more similar to dogs than they are to plants.

We may represent this theory about what makes analogy arguments stronger or weaker in the following way. Object t is the *target*—it is the object about which one aims to draw a conclusion. Object a is the *analog*, which is already known to possess the property P:

Object a has property P.
Object a and object t are similar to degree n.
n ═══════════
Object t has property P.

In this argument skeleton, n occurs twice. It measures the degree of overall similarity between a and t, where $n = 0$ means that the two objects have no properties in common and $n = 1$ means that they have all the same properties. The variable n also measures how strongly the premisses support the conclusion, where this, too (like the concept of probability itself), can have a value anywhere from 0 to 1, inclusive. The more similar the analog and the target, the more strongly the premisses support the conclusion.

Hume believes that this fairly plausible theory about the logic of analogy arguments has important consequences for the design argument. To see how strongly the premisses support the conclusion of the design argument, we must ask how similar watches and organisms really are. A moment's reflection shows that they are very *dis*similar. Watches are made of glass and metal; they do not breathe, grow, excrete, metabolize, or reproduce. The list could go on and on. Indeed, it is hard to think of two things that are more *dis*similar than an organism and a watch. The immediate consequence, of course, is that the design argument is a very weak analogy argument. It is preposterous to infer that organisms have a given property simply because watches happen to have it.

Although Hume's criticism is devastating if the design argument is an argument from analogy, I see no reason why the design argument must be construed in this way. Paley's argument about organisms stands on its own, regardless of whether watches and organisms happen to be similar. The point of talking about watches is to help the reader see that the argument about organisms is compelling.

To drive this point home, consider a third application of the Likelihood Principle. Suppose we toss a coin a thousand times and note on each toss whether the coin lands heads or tails. We record the observational results in statement *O* below and wish to use *O* to discriminate between two competing hypotheses:

O: The coin landed heads on 803 tosses and tails on 197.
H_1: The coin is biased toward heads—its probability of landing heads when tossed is 0.8.
H_2: The coin is fair—its probability of landing heads when tossed is 0.5.

The Likelihood Principle tells us that the observations strongly favor H_1 over H_2. The evidence points toward one hypothesis and away from the other. This is a standard idea you might hear in a statistics class. It is quite irrelevant to this line of reasoning to ask whether the coin is similar to an organism or to a watch or to anything else. Likelihood stands on its own; analogy is irrelevant.

I now turn to Hume's second criticism of the design argument, which is no more successful than the first. He asserts that an inference from an observed effect to its conjectured cause must be based on induction. Suppose we observe that Sally has a rash on her arm. We infer from this that she had contact with poison ivy. Hume insists that this inference from effect to cause is reasonable only if it is based on prior knowledge that such rashes are usually caused by exposure to poison ivy.

What determines whether such an inductive argument is stronger or weaker? If we have examined only a few cases of rashes and have observed that most of them are caused by poison ivy exposure, then it is a rather weak inference to conclude that Sally's rash was produced by poison ivy exposure. On the other hand, if we have looked at a large number of rashes and have found that poison ivy caused all of them, then we would be on much firmer ground in our claim that Sally's rash was due to poison ivy.

Hume's idea corresponds to the modern idea that *sample size* is an important factor in determining whether an inference is strong or weak. Hume thinks that this consideration has devastating implications when it is applied to the design argument. He contends that if we are to have good reason to think that the organisms in *our* world are the product of intelligent design, then we must have looked at lots of *other* worlds and observed intelligent designers producing organisms there. But how many such worlds have we observed? The answer is, *not even one*. So the inductive argument is as weak as it possibly could be; its sample size is zero.

Once again, it is important to see that an inference to the best explanation need not obey the rules that Hume stipulates. For example, consider the suggestion by Alvarez *et al.* (1980) that the mass extinction that occurred at the end of the Cretaceous period was caused by a large meteorite crashing to earth and sending up a giant dust cloud. Although there is plenty of room to disagree about whether this is plausible (see Jablonski 1984 for discussion), it is quite irrelevant that we have never witnessed meteorite strikes producing mass extinctions "in other worlds." Inference to the best explanation is different from an inductive sampling argument.

Hume produced other criticisms of the design argument, but these fare no better than the two I have described here. Part of the problem is that Hume had no serious alternative explanation of the phenomena he discusses. It is not impossible that the design argument should be refutable without anything being provided to stand in its stead. For example, this could happen if the hypothesis of an intelligent designer were incoherent or self-contradictory. But I see no such defect in the argument.

It does not surprise me that intelligent people strongly favored the design hypothesis when the only alternative available to them was random physical processes. But Darwin entirely altered the dialectical landscape of this problem. His hypothesis of evolution by natural selection is a third possibility; it requires no intelligent design, nor is natural selection properly viewed as a "random physical process." Likelihood considerations favor design over randomness, but whether likelihood will favor design over evolution by natural selection remains to be seen.

2.4 Why Natural Selection Isn't a Random Process

Natural selection occurs when there is heritable variation in fitness. An organism's fitness is its ability to survive and reproduce, which is represented in terms of probabilities. For example, suppose the organisms in a population differ in their abilities to survive from the egg to the adult stage. This will mean that different organisms have different probabilities of surviving.

Since fitnesses are represented in terms of probabilities, there is a sense in which chance plays a role in evolution by natural selection. But if chance plays a role, doesn't this mean that natural selection is a random process? And if natural selection is a random process, how can it constitute a form of explanation that differs from the alternatives that Paley considered in his design argument?

If a process is random, then different possibilities have the same (or nearly the same) probabilities. A fair lottery involves random draws from an urn; each ticket has the same chance of winning. However, when the different possibilities have drastically unequal probabilities, the process is not a random one. If I smoke cigarettes, eat fatty foods, and don't exercise, my probability of a long life may be lower than yours if you avoid these vices. In this case, the determination of which of us lives and which of us dies does not proceed at random.

Natural selection involves *un*equal probabilities, and for this reason, it is not a random process. Randomness becomes an issue in the theory of evolution when the neutrality hypothesis is considered. If the alleles present at a locus in a population are equal (or nearly equal) in fitness, then gene frequencies change because of random genetic drift, not because of natural selection. Randomness is an important issue in the theory of evolution, but it is not part of the process of natural selection.

Creationists sometimes describe natural selection as "random" when they compare it to a tornado blowing through a junkyard. The tornado "randomly" rearranges the pieces of junk. It is enormously improbable that this "random" activity should put together a functioning automobile. Creationists think the same is true of natural selection: Because it is "random," it cannot create order from disorder.

It is possible to give this line of thinking the appearance of mathematical precision. Consider the billions of ways the parts in a junkyard might be brought together. Of these many combinations, only a tiny fraction would give rise to a functioning automobile. Therefore, it is a safe bet that a tornado won't have this result. Notice how this argument connects with the definition of randomness given above. Implicit in the argument is the idea that each arrangement of parts is just as probable as any other. Given this assumption, the conclusion really does follow. However, it is a mistake to think that natural selection is a process in which every possible outcome has the same probability.

The process of natural selection has two components. First, variation must arise in the population; then, once that variation is in place, natural selection can go to work, modifying the frequencies of the variants present. Evolutionists sometimes use the word "random" to describe the mutation process but in a sense slightly different from the one I just described. Mutations are said to be "random" in that they do not arise because they would be beneficial to the organisms in which they occur. There may be physical reasons why a given mutagen—radiation, for example—has a higher probability of producing one mutation than some other. "Random mutation" does not mean that the different mutants are equiprobable.

The fact that the mutation-selection process has two parts has an important bearing on the creationist's analogy of the tornado sweeping through the junkyard. It is brought out vividly by Richard Dawkins (1986) in his book *The Blind Watchmaker*. Imagine a device that is something like a combination lock. It is composed of a series of disks placed side by side. On the edge of each disk, the twenty-six letters of the alphabet appear. The disks can be spun separately so that different sequences of letters may appear in a viewing window.

How many different combinations of letters may appear in the window? There are 26 possibilities on each disk and 19 disks in all. So there are 26^{19} different possible sequences. One of these is METHINKSITISAWEASEL. If the disks turn independently of each other and if each entry on a disk has the same chance of appearing in the viewing window, then the probability that METHINKSITISAWEASEL will appear after all the disks are spun is $1/26^{19}$, which is a very small number indeed. If the process is truly random, in the sense just described, then it is enormously improbable that it could produce the orderly message just mentioned, even if the disks were spun repeatedly.

Even with a *billion* spins of all the disks together, the probability of hitting the target message is still vanishingly small.

Now let's consider a quite different process. As before, the disks are spun and they are "fair"; each of the 26 possibilities has the same chance of appearing in the viewing window for that disk. But now imagine that a disk is frozen if it happens to put a letter in the viewing window that matches the one in the target message. The remaining disks that do not match the target then are spun at random, and the process is repeated. What is the chance now that the disks will display the message METHINKSITISAWEASEL after, say, fifty repetitions?

The answer is that this message can be expected to appear after a surprisingly small number of generations of the process. Of course, if we all had such devices and each of us ran the experiment, some of us would reach the target sentence sooner than others. But it is possible to calculate what the average number of generations is for the process to yield METHINKSITISAWEASEL. This average number is not very big at all.

Although the analogy between this process and the mutation-selection process is not perfect in every respect, it does serve to illustrate an important feature of how evolution by natural selection proceeds. *Variation* is generated without regard to whether it "matches the target" (i.e., is advantageous to the organism). But *retention* (selection among the variants that arise) is another matter. Some variants have greater staying power than others.

A wind blowing through a junkyard is, near enough, a random process. So is repeatedly spinning all the disks of the device just described. But the mutation-selection process differs crucially from both. Variation is generated at random, but selection among variants is nonrandom.

2.5 Two Kinds of Similarity

I have argued that the design hypothesis is more likely than the hypothesis of random natural processes when each is asked to explain why organisms are intricate and well adapted. The task now is to see how the design hypothesis compares with the hypothesis of evolution by natural selection. Once again, the analytic tool we will use is the Likelihood Principle.

There is a crucial respect in which the observations that require explanation have "changed" since the time of Paley. Paley stressed the adaptive perfection of nature. He believed that each detail of living things is for the best. Paley was not alone in this respect. A century or so earlier, the philosopher/scientist Gottfried Leibniz (1646–1716) argued that God had brought into being the best of all possible worlds. Voltaire satirized this optimistic idea in his comedy *Candide* through the character of Dr. Pangloss, who stumbles around the world naively seeing perfection in every detail.

Darwin began the break with this perfectionist tradition, and modern evolutionists have followed Darwin's lead. They reject the idea that adaptation is perfect; rather, they argue that adaptation is good enough, imperfect though it almost always is. More precisely, what natural selection predicts is that the fittest of the traits actually represented in a population will become common.

The result is not the best of all conceivable worlds but the best of the variants actually available.

Natural selection is a "tinkerer" (Jacob 1977). Organisms are not designed from scratch by a supertalented engineer. Instead, a present-day organism has traits that are modifications of the traits found in its ancestors. This contrast between the hypothesis of evolution by natural selection and the design hypothesis is of the utmost importance. The two theories make quite different predictions about the living world.

Consider the fact that organisms in various species often exhibit structural differences among parts that perform the same function. Wings in birds, bats, and insects all facilitate flight. Yet, close attention to these "wings" reveals that they differ in numerous respects that have little or nothing to do with the requirements of flight. If wings were designed by an intelligent engineer so that they would optimally adapt the organism for flight, it would be very hard to explain these differences. On the other hand, they become readily intelligible if one accepts the hypothesis that each of these groups is descended from wingless ancestors. The bird's wing is similar to the forelimbs of its wingless ancestors. A bat's wing is likewise similar to the forelimb of its wingless ancestors. Wings were not designed from scratch but are modifications of structures found in ancestors. Because natural selection is a tinkerer, organisms retain characteristics that reveal their ancestry.

A similar line of argument is based on vestigial organs. Human fetuses develop gill slits and then lose them. The embryos of whales and anteaters grow teeth, which then are resorbed into the jaw before birth. These traits are entirely useless to the organism. It is puzzling why an intelligent designer would have inserted them into the developmental sequence only to delete them a short time later. However, these vestiges are not at all puzzling once it is realized that humans, whales, and anteaters each had ancestors in which the traits were retained after birth and had a function. Gill slits lost their advantage somewhere in the lineage leading to us, so they were deleted from the adult phenotype. Their presence in the embryo did no harm, so the embryonic trait has persisted.

Vestigial traits also are found in adult organisms. Why is the human spinal column so similar to the spinal column found in apes? The shape of this common spine is most unsuited to upright gait, but it makes more sense for an organism that walks on all fours. An engineer who wished to equip monkeys with what they need and human beings with what they need would not have provided the same arrangement for each (assuming, that is, that the engineer is benevolent and does not wish to promote back pain). However, if human beings are descended from ape ancestors, the similarity is not surprising at all. The ancestral condition was modified to allow human beings to walk upright, although this modified condition is not perfect in all respects.

Gould (1980b) tells a similar story about the panda's "thumb." Pandas are vegetarians whose main food is bamboo. The panda strips the bamboo by running the branch between its paw and a spur of bone (a "thumb") that juts out from its wrist. This device for preparing food is quite inefficient; an engineer easily could have done better. However, the paw of the panda is remarkably similar to the paws of carnivorous bears. Why are the paws so similar, given that pandas and other bears have such different dietary requirements? Once again, the similarity makes sense as a vestige of history, not as a product of optimal design. Pandas are descended from carnivorous bears, and so their paws are modifications of an ancestral condition.

Evolutionists maintain that facts about biogeographic distribution constitute a third kind of observation that supports evolution by natural selection better than it supports the design hypothesis. Sightless fish live in lightless caves the world over. The physical environments in these caves are remarkably similar. If God had built organisms to suit them to the environments they inhabit and if he gave them designs that optimally suit them to their ways of life, why do these sightless fish differ from each other in so many respects? The explanation is that different sightless fish are descended from different sighted species of fish, which usually live nearby. Again, what does not make sense from the point of view of optimal design becomes readily intelligible as a vestige of the evolutionary past.

Traits of the kinds just described are quite common. It isn't just occasionally that biologists confront a trait that has no adaptive explanation: This situation is absolutely routine. If this were a biology book or a book-length treatment of the evidence for evolution and against creationism (for which see Futuyma 1982 and Kitcher 1982a), I would pile up more data of the kind to which I have just alluded. But since this is a text in philosophy, I want to focus more on the logic of these arguments than on their empirical details.

I hope it is clear how these arguments make use of the Likelihood Principle. Some observation (*O*) is cited, and the design hypothesis (*D*) and the hypothesis of evolution by natural selection (*E*) are considered in its light. The claim is made that the observation would be very surprising if the design hypothesis were true but would be quite unsurprising if the hypothesis of evolution by natural selection were correct. The observation strongly favors evolution over design because $P(O/E) >> P(O/D)$.

I have just cited examples of similarities among species that are evidence of their common ancestry. However, there are other similarities that do not have this status. Humans and apes are both able to extract energy from their environments; in addition, both are able to reproduce. But these similarities do not offer strong evidence for relatedness because we would expect organisms to be able to do these things even if they were unrelated genealogically. Some similarities favor the hypothesis of common ancestry; others do not discriminate between that hypothesis and the hypothesis of separate origination. What distinguishes the one type of similarity from the other?

In replying to the challenge of creationism, biologists often find themselves explaining why natural selection is a very powerful force. If asked how the vertebrate eye could have evolved by natural selection, biologists attempt to show how a sequence of gradual modifications could transform a light-sensitive organ into the camera-like adaptation that we use to see. It is easy to misinterpret such lines of reasoning and conclude that natural selection is inclined to do precisely the same thing that a superintelligent engineer would do. If engineering considerations suggest that a device for seeing must have a lens that focuses incoming light, then natural selection can be expected to produce a lens that does precisely that.

Such claims for the power of natural selection run the risk of obscuring the best evidence there is for the other half of Darwin's two-part theory—the hypothesis of the tree of life. Darwin recognized this important property of his theory in the following passage from the *Origin* (p. 427):

> On my view of characters being of real importance for classification only in so far as they reveal descent, we can clearly understand why analogical or adaptive characters, although of the utmost importance to the welfare of the being, are almost valueless to the systematist. For animals, belonging to two most distinct lines of descent, may readily become adapted to similar conditions, and thus assume a close external resemblance; but such resemblances will not reveal—will rather tend to conceal their blood-relationship to their proper lines of descent.

Adaptive characters are good for the organism, but the adaptive similarities displayed by organisms are bad for the systematist who wishes to reconstruct the genealogies of the species involved. If two species share a feature that has an obvious adaptive rationale, this similarity will be "almost valueless" in defending the claim that the two species have a common ancestor.

I have formulated the tree of life hypothesis as a very strong (i.e., logically ambitious) claim. It says that *all* present-day organisms on earth are genealogically related to each other. But why accept the tree of life hypothesis in this strong form? For example, why not think that animals are related to each other and that plants are related to each other but that plants and animals have no common ancestors?

One standard line of evidence used to answer this question is the (near) universality of the genetic code. This is not the fact that all terrestrial life is based on DNA/RNA. Rather, it concerns the way strands of DNA are used to construct amino acids, which are the building blocks of proteins (and hence of larger-scale developmental outcomes). Messenger RNA consists of sequences of four nucleotides (*A*denine, *C*ytosine, *G*uanine, and *U*racil). Different nucleotide triplets (codons) code for different amino acids. For examine, *UUU* codes for Phenylalanine, *AUA* codes for Isoleucine, and *GCU* codes for Alanine. With some minor exceptions, all living things use the same code. This is interpreted as evidence that all terrestrial life is related.

Biologists believe that the code is *arbitrary*—there is no functional reason why a given codon should code for one amino acid rather than another (Crick 1968). Notice how the arbitrariness of the code plays a crucial role in this likelihood argument. If the code is arbitrary, then the fact that it is universal favors the hypothesis that all life shares a common origin. However, if the code is not arbitrary, the argument changes. If the genetic code we observe happened to be the only (or the most functional) physical possibility, we might expect all living things to use it *even if they originated separately.*

Consider an analogy between the problem of reconstructing biological evolution and the problem of reconstructing the evolution of cultures. Why do historical linguists believe that different human languages are related to each other? Why not think that each arose separately from all the others? It isn't just that languages display a set of similarities. Indeed, there are some similarities between languages that we would expect even if they had originated separately. For example, the fact that French, Italian, and Spanish all *contain names* is not strong evidence that they are related to each other. Names have an obvious functional utility. On the other hand, the fact that these languages assign similar names to numbers is striking evidence indeed:

	French	*Italian*	*Spanish*
1	un	uno	uno
2	deux	due	dos
3	trois	tre	tres
4	quatre	quattro	cuatro
5	cinq	cinque	cinco

To be sure, it is *possible* that each language independently evolved similar names for the numbers. But it is far more *plausible* to suppose that the similarity is due to the fact that the languages share a common ancestor (Latin). Once again, the reason this similarity is such strong evidence for a common ancestor is that the names for given numbers are chosen arbitrarily. Arbitrary similarity, not adaptive similarity, provides powerful evidence of genealogical relationship.

2.6 The Problem of Predictive Equivalence

Why did Paley think that the existence of God was such a plausible explanation of the characteristics of life forms? The answer to this question has two parts. He thought that *organisms* are perfectly adapted to the lives they lead. And he thought that *God* was a perfectly benevolent, knowledgeable, and powerful being. These conceptions of *God* and *life* fit together hand in glove: Change one but not the other and the fit between the design hypothesis and the observations deteriorates. This is what happens when contemporary biology corrects what Paley took to be his observations. If species are *imperfectly* adapted—if many of their traits are makeshift and even useless—then postulating the engineering God that Paley envisioned becomes much less plausible.

In this section, I want to explore a second sort of modification that might be made to Paley's design hypothesis and his conception of the observational data. Suppose we modified our picture of what God would be like if he existed. There are many ways to do this, and I certainly will not attempt to survey them all. But there are two modifications that are worth considering if we wish to understand the logic of the problem posed by the debate between creationism and evolution.

One possible modification of Paley's picture of God involves removing God from the problem of the origin of species. Suppose one believed that God created the universe and then sat back and let physical laws play themselves out. This version of theism does not conflict with the hypothesis of evolution by natural selection. Of course, whether this version of theism is plausible is a separate matter. One must ask whether there is any reason to think that this sort of being really exists. To explore this question would take us away from the design argument and into more general philosophical issues. The point to notice here is that this version of theism is not a competing hypothesis if one is trying to assess whether the theory of evolution by natural selection is plausible. Evolutionary theory competes with this version of theism no more than it competes with the theory of special relativity; the theories are about totally different phenomena.

The other change in Paley's conception of God that I want to consider appears to undercut the likelihood arguments described in the previous section. Consider the hypothesis that God created each species separately but did so in a way that misleads us into thinking that species evolved by natural selection. This hypothesis of a "trickster" God disagrees with the hypothesis of evolution by natural selection. Yet, the two hypotheses make the same predictions about what we observe in the living world. Because these hypotheses are *predictively equivalent,* no likelihood argument can be used to show that the observations favor one of them over the other. By changing our conception of God from the benevolent engineer envisioned by Paley to the trickster just described, we have rescued the design hypothesis from disconfirmation. Does this mean that the design hypothesis is alive and well and that the hypothesis of evolution by natural selection is not strongly supported by what we know?

Let us schematize this problem to make its logic explicit. We considered the likelihoods of two hypotheses, relative to an observation:

O: Organisms are *im*perfectly adapted to their environments.
D_p: Species were separately created by a superintelligent, benevolent, and omnipotent God.
Ev: Species evolved from common ancestors by the process of natural selection.

The observations favor the hypothesis of evolution (*Ev*) over Paley's design hypothesis (D_p); $P(O/Ev) >> P(O/D_p)$. But now consider a new version of the design hypothesis:

D_t: Species were separately created by a God who made them look just the way they would if they had evolved from common ancestors by the process of natural selection.

This trickster hypothesis is a wild card. D_t and *Ev* are predictively equivalent; whatever evolution by natural selection predicts about the imperfection of organisms, the hypothesis of a trickster God predicts the same thing. The hypotheses therefore have *equal* likelihoods; $P(O/Ev) = P(O/D_t)$. The question we need to consider is whether this fact should weaken our confidence that the evolution hypothesis is true.

When I explained the Likelihood Principle in Section 2.2, I emphasized that a hypothesis with high likelihood might nonetheless be quite implausible. For example, the hypothesis that there are gremlins bowling in the attic has a very high likelihood, relative to the noise we hear, but that does not mean that the noise tells us that the gremlin hypothesis is probably correct. In this case, we have antecedent reasons to regard the existence of gremlins as very implausible. Because of this, the noise in the attic does not and should not convince us that there really are gremlins up there bowling.

Can we offer a comparable argument against the trickster hypothesis D_t? Although this hypothesis has the same likelihood as the hypothesis of evolution by natural selection, are there other reasons why we should dismiss it as implausible? D_t does express a rather unusual conception of what God would be like if he existed. Can we argue that it is quite implausible that God, if he existed, would be a trickster? Perhaps we should go so far as to insist that God is *by definition* perfectly benevolent, knowledgeable, and powerful.

I do not find this suggestion very persuasive. True, it is not common to conceive of God as a trickster, but that is no reason to think that he isn't one. Moreover, I do not think that the definition of the concept of God can be used to settle this question. Different religions conceive of God in different ways, and it is parochial to assume that God must be just the way one religious tradition says he is. Thus, I see no contradiction in the idea that God is a trickster.

A second criticism that might be made of the hypothesis D_t is that it is untestable. The suggestion is that the problem with D_t is not that the evidence makes it implausible but that there is no way to find out if it is plausible. We will explore this suggestion at greater length in the next section. For now, the point is that there is a certain symmetry between the evolution hypothesis and the trickster hypothesis. If it is claimed that the trickster hypothesis is untestable, won't the same be true of the evolution hypothesis? After all, these two theories make the same predictions.

The problem we face here derives from the fact that the Likelihood Principle is a comparative principle. We test a hypothesis by testing it *against one or more competing hypotheses.* The observations favor *Ev* over D_p, but they do not favor *Ev* over D_t. As far as likelihood is concerned, the evolution hypothesis and the trickster hypothesis are in the same boat.

Although likelihood does not discriminate between the evolution hypothesis and the trickster hypothesis, I suggest that this is no reason to doubt the truth of the theory of evolution. The predictive equivalence of *Ev* and D_t demonstrates no special defect in *Ev*. Take any of the perfectly plausible beliefs you have about the world. It is possible to construct a trickster alternative to that plausible belief in such a way that the plausible belief and the trickster alternative are predictively equivalent.

Consider, for example, your belief right now that there is a printed page in front of you. Why do you think this is true? The evidence you have for this belief derives from the visual and (perhaps) tactile experiences you now are having. This evidence strongly favors the hypothesis that there is a printed page in front of you, as opposed to, say, the hypothesis that there is a baseball bat there. That is, $P(O/Page) \gg P(O/BB)$, where

O: your present sensory experiences
Page: There is a printed page in front of you.
BB: There is a baseball bat in front of you.

It is common sense to think that the experiences you now are having give you a strong reason to think that *Page* is true but very little reason to think that *BB* is true. The Likelihood Principle describes why this makes sense.

However, now let us introduce a wild card (inspired by the evil demon of Descartes's *Meditations*). It is the trickster hypothesis:

Trick: There is no printed page in front of you; however, a trickster God is causing you to have precisely the experiences you would be having if there were a printed page in front of you.

Although likelihood favors *Page* over *BB,* likelihood does not favor *Page* over *Trick.* The reason is that *Page* and *Trick* are predictively equivalent.

How should you interpret the fact that *Page* is not more likely that *Trick*? Perhaps you feel that you can muster considerations that explain why *Page* is more plausible than *Trick* even though the two hypotheses are predictively equivalent. Perhaps you are skeptical that this can be done. I don't want to address which of these attitudes is defensible. My point is to note a structural similarity between the three explanations of your current visual impressions and the three explanations discussed before of the adaptive imperfection of organisms.

Even beliefs that you think are *obviously* true (like there being a printed page before you now) can be confronted with the problem of predictive equivalence. *Page* is something you may think is clearly right, but it is hard to see how to discriminate between it and *Trick.* The fact that the hypothesis of evolution by natural selection faces the same problem, therefore, does not show that there is anything especially weak or dubious about it.

I have considered two possible versions that the design hypothesis might take. Both claim that species were separately created by God; they disagree about what God would be like if he existed. Paley's God is a perfecting being; the trickster God conceals his handiwork. Of course, there are more than two possible conceptions of what God the designer of organisms would be like. This means that there are many more versions of the design hypothesis than the two I have surveyed.

Paley's version of the design hypothesis is undermined by what we observe in nature. The same cannot be said of the trickster version of the design hypothesis. I have argued that if you believe there is a printed page in front of you and reject the trickster explanation of the visual experiences you now are having, you also should not take seriously the trickster version of the design hypothesis. The possibility remains, however, that there is some other version of the design hypothesis that both disagrees with the hypothesis of evolution and also is a more likely explanation of what we observe. No one, to my knowledge, has developed such a version of the design hypothesis. But this does not mean that no one ever will.

2.7 Is the Design Hypothesis Unscientific?

In discussing Paley's version of the design argument, I have emphasized that it can be tested like any scientific hypothesis and that, when tested, it is found wanting. In arguing this way, I have followed the lead of many biologists, who have taken pains to point out how the hypothesis of evolution by natural selection makes predictions that differ dramatically from those that flow from the design hypothesis (i.e., from the version of the design hypothesis they consider).

At the same time and often in the same book, some biologists and philosophers have pursued a quite different line of attack. They have argued that creationism is not a scientific hypothesis because it is untestable. It should be clear that this line of criticism is not compatible with the likelihood arguments we have reviewed. If creationism cannot be tested, then what was one doing when one emphasized the imperfection of nature? Surely it is not possible to test and find wanting a hypothesis that is, in fact, untestable.

The charge of untestability is often developed by appeal to the views of Karl Popper (1959, 1963), who argued that falsifiability is the hallmark of a scientific statement. In this section, I will discuss Popper's ideas. My goal is to develop criticisms of his position and also to reach some wider assessment of the merits of testability as an appropriate criterion for scientific discourse.

As a preliminary point, recall a distinction that was discussed at the beginning of this chapter. When we consider whether something is scientific, we must be clear about whether we are talking about *people* or *propositions.* If someone behaves dogmatically, refusing to look at relevant evidence, then that person has adopted an unscientific attitude. But it does not follow that the proposition the person believes is unscientific. Flat-earthers may take a quite unscientific attitude toward the proposition *the earth is flat;* it does not follow that the proposition is unscientific—i.e., that it cannot be tested.

The relevance of this point to the creationism controversy should be obvious. Creationists often distort scientific findings. They trot out the same old tired arguments, even after these have been refuted repeatedly. They do so without acknowledging that the arguments have been challenged on scientific grounds. I think there is little doubt that most creationists have behaved in a patently unscientific manner. However, it does not follow that creationist *theories* are unscientific. If the theories are unscientific, some further argument must be produced to show that they are.

Popper's basic idea is that scientific ideas are *falsifiable;* they *stick their necks out,* whereas unscientific ideas do not. Less metaphorically, scientific propositions make predictions that can be checked observationally. They make claims about the world that, at least in principle, are capable of conflicting with what we observe. Unscientific ideas, on the other hand, are compatible with all possible observations. No matter what we observe, we can always retain our belief in an unscientific proposition.

Do not confuse falsifiability with actual falsehood. Many true propositions are falsifiable. Indeed, a scientific proposition should run the risk of refutation. But if it actually is refuted, we no longer retain it in the corpus of what we believe. According to Popper, our beliefs should be *falsifiable,* not *false.*

Popper thinks that propositions that express religious convictions about God are unfalsifiable. If you think that God created the living world, you can hold on to that belief no matter what you observe. If you start by thinking of God in the way that Paley does, the observation of imperfect adaptations may lead you to change your mind. However, instead of abandoning your opinion that God created the world, you can modify your picture of what God wanted to do. In fact, no matter what you observe in nature, some version or other of theism can be formulated that is compatible with those observations. Theism, therefore, is unfalsifiable.

Popper also thinks that Freudian psychoanalytic theory is unfalsifiable. No matter what the patient says, the psychoanalyst can interpret the patient's behavior so that it is compatible with psychoanalytic ideas. If the patient admits that he hates his father, that confirms the Freudian idea of the Oedipus complex. If he denies that he hates his father, that shows that he is repressing his Oedipal fantasies because they are too threatening.

Popper has the same low opinion of Marxism. No matter what happens in capitalist societies, the Marxist can interpret those events so that they are compatible with Marxist theory. If a capitalist society is beset by fiscal crisis, that shows that capitalism is collapsing under the weight of its internal contradictions. If the society does not experience crashes, this must be because the working class has yet to be sufficiently mobilized or because the rate of profit has not fallen far enough.

In fact, at one time, Popper also thought that the theory of evolution is not a genuine scientific theory but, instead, is a "metaphysical research program" (Popper 1974). Once again, the idea is that the convinced evolutionist can interpret the observations so that they are consistent with evolutionary theory, no matter what those observations turn out to be. Popper subsequently changed his mind about this.

Notice that, in the previous four paragraphs, I illustrated what Popper has in mind by talking about *people.* I said that the committed theist, the committed psychoanalyst, the committed Marxist, and the committed evolutionist can interpret what they observe so as to hold on to their pet theories. It should be clear by now that this, *per se,* shows us nothing about the *propositions* that figure in those theories. Sufficiently dogmatic people can hold on to any proposition at all, but that does not tell us whether the proposition in question is testable. We now must leave this informal statement of Popper's idea behind and examine what his criterion is for a proposition to be falsifiable.

Popper's criterion of falsifiability requires that we be able to single out a special class of sentences and call them *observation sentences.* A proposition is then said to be falsifiable precisely when it is related to observation sentences in a special way:

> Proposition *P* is falsifiable if and only if *P* deductively implies at least one observation sentence *O*.

Falsifiable propositions make predictions about what can be checked observationally; this idea is made precise by the idea that there is a deductive implication relation between the proposition *P* and some observational report *O*.

One problem with Popper's proposal is that it requires that the distinction between observation statements and other statements be made precise. How might this be done? If an observation statement is one that a person can check without knowing anything at all about the world, then there probably are no observation statements. To check the statement "The chicken is dead," you must know what a chicken is and what death is. This problem is sometimes expressed by saying that *observation is theory laden*. Every claim that people make about what they observe depends for its justification on their possessing prior information.

Popper addresses this problem by saying that what one regards as an observation statement is a matter of convention. But this solution will hardly help one tell, in a problematic case, whether a statement is falsifiable. If one adopts the convention that "God is the creator of the universe" is an observation statement, then theism becomes a falsifiable position. For Popper's criterion to have some bite, there must be a *non*arbitrary way to distinguish observation sentences from the rest. To date, no one has managed to do this in a satisfactory manner.

The problems with Popper's falsifiability criterion go deeper. First, there is the so-called *tacking problem*. Suppose that some proposition *S* is falsifiable. It immediately follows that the conjunction of *S* and any other proposition *N* is falsifiable as well. That is, if *S* makes predictions that can be checked observationally, so does the conjunction *S&N*. This is an embarrassment to Popper's proposal since he wanted that proposal to separate nonscientific propositions *N* from properly scientific propositions *S*. Presumably, if *N* is not scientifically respectable, neither is *S&N*. The falsifiability criterion does not obey this plausible requirement.

Another problem with Popper's proposal is that it has peculiar implications about the relation of a proposition to its negation. Consider a statement of the form "All *A*s are *B*." Popper judges this statement falsifiable since it would be falsified by observing a single *A* that fails to be *B*. But now consider the negation of the generalization—the statement that says "There exists an object that is both *A* and not-*B*." This statement is *not* falsifiable; no single observed object or finite collection of them can falsify this existence claim. So the generalization is falsifiable, though its negation is not. But this is very odd—presumably, if a statement is "scientific," so is its negation. This suggests that falsifiability is not a good criterion for being scientific.

Still another problem with Popper's proposal is that most theoretical statements in science do not, all by themselves, make predictions about what can be checked observationally. Theories make testable predictions only when they are conjoined with auxiliary assumptions. Typically, *T* does not deductively imply *O*; rather, it is *T&A* that deductively implies *O* (here, *T* is a theory,

O is an observation statement, and A is a set of auxiliary assumptions). This idea is sometimes called *Duhem's Thesis;* it is named for Pierre Duhem (1861–1916), a physicist, historian, and philosopher of science who noted this pervasive pattern in physical theories.

Here's an example that illustrates Duhem's Thesis. Consider the hypothesis mentioned earlier that the Cretaceous extinction was caused by a meteor colliding with the earth. Alvarez *et al.* (1980) argued that the presence of the metal iridium in various geological deposits was strong evidence for this hypothesis. The point here is that the meteor hypothesis, by itself, says nothing about where iridium should be found. The argument requires further assumptions—for example, that iridium concentrations are higher in meteors than they are on earth and that deposits from a meteor hit should be found in certain geological strata. Thus, the meteor hypothesis predicts what we observe only when it is supplemented with further assumptions.

The final problem with Popper's proposal is that it entails that probability statements in science are unfalsifiable. Consider the statement that a coin is fair—that its probability of landing heads when tossed is 0.5. What can be deduced from this statement about the observable behavior of the coin? Can one deduce that the coin must land heads precisely five times and tails precisely five times if it is tossed ten times? No. The hypothesis that the coin is fair is logically compatible with all possible outcomes: It is possible for a fair coin to land heads on all ten tosses, to land heads on nine and tails on one, and so on. Probability statements are not falsifiable in Popper's sense.

This does not mean that probability statements are not testable in some reasonable sense of that word. We have already discussed the Likelihood Principle, which plays a crucial role in assessing how observations bear on competing hypotheses. In fact, something like the Likelihood Principle is what Popper himself adopted when he recognized that probability statements are not falsifiable. This objection to the falsifiability criterion was one that Popper himself anticipated.

Where does this leave the question of whether creationism is testable? If Popper's criterion is not a plausible one to use in evaluating science itself, it should not be used to evaluate creationism. Perhaps "God exists" does not, by itself, deductively imply any observation statements. But the same is true for many scientific theories. Furthermore, once "God exists" is conjoined with auxiliary assumptions, it may emerge that the conjunction does issue in predictions that can be checked observationally. And as we vary the auxiliary assumptions, the theistic hypothesis makes different predictions about observations.

We may schematize this point as follows. Consider these three statements:

G: God separately created living things.
A_p: If God had separately created living things, he would have made them perfectly adapted to their environments.
A_t: If God had separately created living things, he would have made them look just the way they would have looked if they had been the product of evolution by natural selection.

Box 2.1 Popper's Asymmetry

Popper (1959, 1963) held that there is an asymmetry between falsification and verification. He maintained that it is possible to prove theories false but impossible to prove them true. Let *T* be a theory and *O* be an observation statement deducible from *T*. If the predicted observation turns out to be false, we can deduce that the theory is false; but if the predicted observation statement turns out to be true, we cannot validly deduce that the theory is true. Popper maintained that there is an asymmetry between falsification and verification because one but not the other of the following arguments is deductively valid:

Falsification	*Verification*
If *T*, then *O*	If *T*, then *O*
not-*O*	*O*
not-*T*	*T*
deductively valid	deductively invalid

Popper took the difference between these two arguments to support a kind of *skepticism*. He held that it is impossible to know that a theory is true; science can tell us only that a theory is false (or that it has yet to be refuted).

If we represent the role that auxiliary assumptions (*A*) play in testing (Duhem's Thesis), symmetry can be restored:

Falsification	*Verification*
If *T*&*A*, then *O*	If *T*&*A*, then *O*
not-*O*	*O*
not-*T*	*T*
deductively invalid	deductively invalid

continues

G, by itself, entails no observation statements. But the conjunction $G\&A_p$ has numerous observational implications; the same is true of $G\&A_t$. This is why we were able to argue that Paley's theory $G\&A_p$ is poorly supported by the evidence but that the conjunction $G\&A_t$ agrees with evolutionary theory with regard to what it predicts about the adaptedness of organisms.

Popper's concept of falsifiability is not much help here, so let us set it aside. Is there a plausible account of scientific testability that we can use to evaluate whether creationism is testable?

Let us take to heart the idea that theoretical statements can be brought into contact with observable phenomena only through the mediation of auxiliary assumptions. The question then arises of how one is to choose those assump-

Box 2.1 continued

If the prediction turns out to be false, we can deduce only that something is wrong in the conjunction *T&A;* whether the culprit is *T* or *A* (or both) remains to be determined.

A vestige of Popper's asymmetry can be restored if we include the premiss that the auxiliary assumptions (*A*) are true:

Falsification	*Verification*
If *T&A*, then *O*	If *T&A*, then *O*
A	*A*
not-*O*	*O*
not-*T*	*T*
deductively valid	deductively invalid

Although we now seem to have a difference between verification and falsification, it is important to notice that the argument falsifying *T* requires that we be able to assert that the auxiliary assumptions *A* are *true*. Auxiliary assumptions are often highly theoretical; if we can't *verify A*, we will not be able to *falsify T* by using the deductively valid argument form just described.

In the last pair of displayed arguments, one is deductively valid and the other is not. However, this does nothing to support Popper's asymmetry thesis. In fact, we should draw precisely the opposite conclusion: The left-hand argument suggests that *if we cannot verify theoretical statements, then we cannot falsify them either*.

One problem with Popper's asymmetry thesis is that it equates *what can be known* with *what can be deduced validly from observation statements*. However, science often makes use of *non*deductive argumentation, in which the conclusion is said to be rendered plausible or to be well supported by the premisses. In such arguments, the premisses do not absolutely guarantee that the conclusion must be true. Perhaps if we reject Popper's deductivism, we can defend an account of scientific inference in which theories can be confirmed *and* disconfirmed. This would establish a symmetry thesis quite alien to Popper's outlook.

tions. That those auxiliary assumptions should be subject to independent check is an appropriate scientific goal.

Consider, for example, a problem that Sherlock Holmes might face. He goes to the scene of a crime and observes various clues. There, he sees the murder victim, killed by a bullet. A number of footprints and a peculiar cigar ash lie nearby. Holmes wishes to figure out who the murderer was. One possibility is that Moriarty is the culprit. How might Holmes determine whether this hypothesis is well supported by the available evidence?

Of course, he cannot directly observe the crime; it is now past. Rather, what Holmes can do is test hypothesis *M* by seeing what *M* predicts about the clues *O* he has observed:

M: Moriarty is the murderer.
O: The victim was killed by a bullet, and there were largish footprints and an ash from an El Supremo cigar at the scene of the crime.

Note that *M* does not, by itself, make predictions about whether *O* should be true. What is needed is further information. Consider two possible sets of auxiliary assumptions:

A_1: Moriarty's preferred weapon is a gun, he has large feet, and he smokes El Supremos.
A_2: Moriarty's preferred weapon is a knife, he has small feet, and he does not smoke.

Note that the conjunction $M\&A_1$ predicts that *O* will be true but that $M\&A_2$ does not. Different auxiliary assumptions lead the hypothesis *M* to make different predictions about the observations.

Holmes's next step will be to figure out whether A_1 or A_2 is true. He should not accept either on faith but should seek further evidence that helps decide the matter. Suppose that, for some reason, Holmes cannot find out which set of auxiliary assumptions is correct. If so, his investigation of Moriarty will be stopped dead.

I believe that creationism faces an analogous problem. The hypothesis that God separately created living things is testable only when it is conjoined with auxiliary assumptions. But how is one to know which auxiliary assumptions to believe? Paley saw God one way; other creationists may prefer a different picture of what God would be like. Different religions conceive of God in different ways. And there are conceptions of God (like the trickster God discussed before) that perhaps are not part of any mainstream religion. How is one to choose? The fact that some of these conceptions of God are familiar while others are decidedly odd is no basis for selecting. What one wants is evidence that one of them is *true* and that the rest are *false*. Without any evidence of this sort, the project of testing the hypothesis that God separately created the species that populate the living world is stopped dead.

I do not claim that no one will ever be able to formulate an argument that shows which auxiliary hypothesis about God is correct. I do not claim to be omniscient. But, to date, I do not think that this issue has been resolved satisfactorily. Perhaps one day, creationism will be formulated in such a way that the auxiliary assumptions it adopts are independently supported. My claim is that no creationist has succeeded in doing this yet.

Let me summarize how my assessment differs from the analysis suggested by Popper. A Popperian criticism of creationism will claim that the theory makes no predictions. I do not think that this criticism is correct since there are numerous versions of creationism—Paley's and the trickster God hypothesis, to mention just two—that do just that. Rather, my present objection is that there is no evidence that allows one to choose between various candidate auxiliary hypotheses.

Box 2.2 The Virtue of Vulnerability

Popper's falsifiability criterion was intended to make precise the idea that scientific hypotheses should be vulnerable to observational test. The falsifiability criterion, we have seen, is unsatisfactory. However, this leaves open the questions of whether a scientific hypothesis ought to be vulnerable to observations and of how this idea of "vulnerability" ought to be characterized.

On the face of it, vulnerability appears to be a defect, not a virtue. Why is it desirable that the hypotheses we believe should be refutable? Wouldn't science be more secure if it were *in*vulnerable to empirical disconfirmation?

The Likelihood Principle helps answer these questions. A consequence of this principle is that *if* O *favors* H_1 *over* H_2, *then not*-O *would favor* H_2 *over* H_1. This is because if $P(O/H_1) > P(O/H_2)$, then $P(\text{not-}O/H_1) < P(\text{not-}O/H_2)$.

We want our beliefs to be supported by observational evidence. For this to be possible, they must be vulnerable; there must be possible observations that would count against them. This requirement is not a vestige of the discredited falsifiability criterion. It flows from the Likelihood Principle itself.

Another difference between my assessment and Popper's is that his is far more ambitious. He thought that the logical properties of a statement settle, once and for all, whether it is testable. My conclusion is more provisional. Thus far, no argument has been stated that allows one to know which auxiliary assumptions should be adopted. Perhaps this will change, but until that happens, creationism cannot be tested.

Creationism, I have emphasized, is a flexible doctrine. It can be developed in many different ways. Right now, the most vocal group is the young earth creationists, who hold views about geology that conflict with a good deal of physics. However, there also are creationists who admit that the earth is quite old and that evolution by natural selection has been very important in the history of life; they nonetheless maintain that God has occasionally intervened at crucial points in the history of life. And of course, there are many more possible forms that creationism might take than the ones that actually find adherents here and now.

In this respect, creationism is like other *isms*. Marxism and Freudianism, to mention two of the examples that Popper singles out for criticism, can be developed in various ways. Some specific formulations have come into conflict with what we observe. Others have not. When one specific formulation is refuted by evidence, another can be constructed in its stead.

This phenomenon is not alien to science but is part of the ongoing project that science pursues. I mentioned at the beginning of this chapter that phrenology was a research program that eventually was cast aside. In Chapter 5, we will consider adaptationism, which is a research program that is both influential and controversial in mainstream evolutionary theory today. The program emphasizes the importance of natural selection in explaining the phenotypic traits we observe. Specific adaptationist explanations of specific traits in specific populations may turn out to be wrong, but that will not stop adaptationists from trying to think up new adaptationist explanations of those traits. Nor should it. Adaptationism is an idea that can be assessed only in the long term, after numerous specific explanations have been developed and tested.

Having stressed this similarity between creationism and the *isms* that animate scientific research, I want to note this difference: It would be a distortion to call creationism a research program because creationists really do not do research at all. Mainly, they try to invent tricky arguments that are intended to convince the naive and throw the unwary off balance. Here, I am talking about people, not propositions. Creation *theory* may be thought of as an idea that can be evaluated in the long term by seeing whether it can solve problems that rival research programs are unable to address. There are many versions that the theory can take; only by exploring a good number of these can we reach a fair assessment of the theory's plausibility. A single set of observations may impugn one version of creationism, but to give the idea a real run for its money, other versions must be explored.

If we view creationism as a flexible doctrine that can be formulated in numerous specific ways, how should we evaluate it? The long-term track record of "scientific creationism" has been poor. Phrenology eventually was discarded; although it showed some promise initially, it failed to progress in the long run. Creationism has fared no better; indeed, it has done much worse. It was in its heyday with Paley, but since then, the idea has moved to the fringe of serious thought and beyond. Perhaps time enough has passed for it to be discarded on the rubbish heap of history.

2.8 The Incompleteness of Science

There are a number of stock arguments that creationists trot out, despite the fact that they have been answered competently and repeatedly by scientists. Many of these have very little philosophical interest; they rest on misunderstandings of scientific ideas that are not hard to correct. For example, creationists complain about the techniques used to date fossils and the geological strata in which they are found. They also maintain that the Second Law of Thermodynamics shows that natural processes cannot generate order from disorder. The interested reader should consult Futuyma (1982) and Kitcher (1982a) for explanations of where creationists go wrong.

However, there is one further sort of argument, with a peculiarly philosophical cast, that helps give creationism its perennial appeal. This argument begins with the fact that there are many features of the living world that evolutionary biology cannot now explain. The origin of life, for instance, remains an active area of scientific research. Earlier, I mentioned the evolution of sex as an important unsolved problem. I also noted that the biological understanding of development—of how a fertilized egg develops into a differentiated organism—is very far from complete. Science is shot through with ignorance. Doesn't this provide an opportunity for creationist explanations to be pressed home?

Scientists do not think they now have all the answers. That is why they continue to do research. On the other hand, creationists have at hand an all-purpose explanation for any observation you please. The origin of life, the distribution of modes of reproduction, and everything else can be explained by a four-word hypothesis: "It was God's will."

We face here a puzzle different from the ones posed by Paley's explanation of adaptation and by the hypothesis of a trickster God. Paley's hypothesis has a lower likelihood than the hypothesis of evolution by natural selection; the trickster hypothesis has the same likelihood as the evolution hypothesis on which it is parasitic. But in the present case, there is no naturalistic hypothesis we can state for the observations of interest. The theistic explanation is *the only game in town.* "It was God's will," if true, would explain the phenomena just listed. No naturalistic competitor can now be formulated as its rival.

When I first introduced the Likelihood Principle, I emphasized that there is more to the plausibility of a hypothesis than its likelihood. If there were gremlins bowling in the attic, that would explain why you now hear noises. Perhaps no other explanation comes to mind. But that does not compel you to believe in gremlins. An alternative response is to admit that you currently have no plausible explanation for the noises you hear.

Creationists try to parlay the current incompleteness of scientific knowledge into points in their favor. Paley believed that the fact of adaptation cannot be explained by known natural processes but requires the hypothesis of a benevolent deity. Creationists nowadays have their own favored stock of examples, some of which really do involve phenomena that current science is unable to address. We can expect creationists in the future to choose a different array of phenomena since many of the problems that currently puzzle science probably will be sorted out in the future.

When creationists use the current incompleteness of evolutionary theory to argue for their position, they should be asked this question: Why do you think that no scientific explanation will ever be developed for the phenomenon in question? Our current ignorance is no evidence for the truth of *any* explanation, creationist or otherwise. The fact that we currently do not understand various facts about life is no reason to think that God has intervened in life's history.

At the same time, the past successes of scientific explanation suggest that what now is inexplicable may eventually be brought within the scope of scientific understanding. Significant advances already have been achieved in understanding the origin of life, the evolution of sex, and the process of ontogeny. Many questions remain to be answered. Can science do the job? It is not utterly crazy to think it can. The track record of science offers grounds for optimism; nothing of the kind can be said concerning the track record of "creation science."

I began this chapter by urging that we not confuse the *propositions* of current creationism and evolutionary theory and the *people* espousing them. To this pair of terms I added a third: I commented on the *research traditions* to which creationism and evolutionary biology both belong. Inquiry involves the testing and modification of theories. Evolutionists change their minds about specific hypotheses and still remain evolutionists; the general framework of evolutionary ideas leaves much room for refinement and debate. Creationism likewise admits of many versions, which may differ in their strengths and weaknesses. A research tradition embodies a general *approach* to a set of problems, which it attempts to address by using a variety of characteristic *techniques.* Research traditions are tested in the long run by seeing if they progress (Laudan 1977; Lakatos 1978): Do problems get solved, or do theories come and go with no net gain in understanding? It is important to realize that creationism is defective not only in its current theories but in its historical track record: Its current theories are unsuccessful *and* its long-term track record has been dismal. It is no surprise that biologists have come to regard "creation science" as a contradiction in terms.

Suggestions for Further Reading

Futuyma (1982) and Kitcher (1982a) defend evolutionary theory and criticize creationism; Gish (1979) and Morris (1974) do the reverse. Ruse (1988a) contains useful philosophical, historical, and scientific essays, and Numbers (1992) provides a thought-provoking history of creationism in the United States.

3

Fitness

A population evolves under natural selection when it contains heritable variation in fitness (Section 1.3). In the present chapter, we will delve more deeply into the fitness concept. What mathematical role does it play in selection models? Since it uses the concept of probability, we must ask how this concept should be understood. And how is the fitness of an organism related to the physical properties the organism and its environment possess? Finally, because the fitness concept is closely related to concepts like advantageousness, adaptation, and function, connections among these concepts must be mapped.

3.1 An Idealized Life Cycle

Natural selection causes a population of organisms to evolve by acting on one or both parts of the organisms' life cycle. Organisms grow from the egg stage (zygotes) to the adult stage; then they reproduce, creating the next generation of zygotes. Organisms may have different probabilities of reaching adulthood, and once they reach adulthood, they may enjoy different degrees of reproductive success. Natural selection occurs when organisms differ in their *viability* and also when they differ in their *fertility*.

Consider a simple case of selection on viability. Suppose that every organism in a population has either trait A or trait B. At the zygote stage in a given generation, these traits occur with frequencies p and q (where $p + q = 1$). Individuals have different probabilities of surviving to adulthood, depending on whether they have A or B. Let us call these different probabilities w_1 and w_2. Note that w_1 and w_2 need not sum to unity. Perhaps A individuals have a 0.9 chance of surviving and B individuals have a chance of surviving of 0.8. How can the zygotic frequencies and the fitnesses of the two traits be used to compute the frequencies that the traits will have at the adult stage? The simple algebraic relationship between the frequencies before selection and after selection is depicted in the following table:

	Traits	
	A	*B*
zygotic frequencies (before selection)	p	q
fitnesses	w_1	w_2
adult frequencies (after selection)	$pw_1/\overline{w}$	$qw_2/\overline{w}$

The quantity $\overline{w}$ (pronounced "w-bar") is the average fitness of the organisms in the population; $\overline{w} = pw_1 + qw_2$. This quantity is introduced to ensure that the frequencies after selection sum to 1; note that $pw_1/\overline{w} + qw_2/\overline{w} = 1$.

A consequence of this algebraic representation of fitnesses is that *fitter traits increase in frequency and less fit traits decline*. If $w_1 > w_2$, then $pw_1/\overline{w} > p$. This must be true since $pw_1/\overline{w} > p$ precisely when $w_1 > \overline{w}$, which simplifies to $w_1 > pw_1 + qw_2$, which must be true if $w_1 > w_2$.

This simple model of viability selection describes what happens in the population during a single generation. What are the long-term consequences of this kind of selection? Let us suppose that *A* and *B* adults have the same number of offspring and that the resulting zygotes are subject to the same viability selection that their parents experienced. Successive generations thus encounter the same selection pressure. Given these assumptions, the greater viability of *A* individuals leads *A* to increase in frequency until it eventually reaches 100 percent (fixation).

In this example, there is differential viability but no differential fertility. The opposite sort of selection process also is possible. Imagine that *A* and *B* individuals have the same probabilities of reaching adulthood but that one type of adult tends to have more offspring than the other. If the organisms reproduce uniparentally, we can use the formalism just described, but we need to reinterpret the fitnesses. When selection acts on viabilities, the fitnesses are probabilities of surviving. Probabilities are numbers between 0 and 1. When there is fertility selection, the fitnesses are *expected numbers of offspring*. These fitnesses need not fall between 0 and 1.

What does "expected" mean? If you toss a fair coin ten times, how many times will it land heads? Clearly, the coin need not land heads exactly five times. But suppose you repeatedly perform this experiment and compute *the average number of heads* obtained in different runs of ten tosses. This average defines the mathematical expectation.

Often, the expected value of some quantity is not the value you would expect to get in any one experiment. The expected number of children in a family in the United States today is about 2.1, but no couple expects to have exactly 2.1 children. The expected value is 2.1 only in the sense that this is the average.

An organism that reproduces uniparentally has different probabilities of producing different numbers of offspring. Suppose that p_i is the probability that a parent will have exactly i offspring ($i = 0,1,2, \ldots$). Then the organism's expected number of offspring is given by the summation $\Sigma i p_i$.

Let the A and B organisms reproduce uniparentally, with e_1 and e_2 as their expected numbers of offspring. I'll assume that offspring always resemble their parents. If adults die after reproducing, then fertility selection will modify population frequencies in the following way:

	Traits	
	A	B
adult frequencies (before selection)	p	q
fitnesses	e_1	e_2
zygotic frequencies (after selection)	$pe_1/\overline{w}$	$qe_2/\overline{w}$

Fertility selection acts on the transition from adult to zygote, not on the transition from zygote to adult.

The simple model just described would have to be adjusted in several ways if the organisms reproduce *bi*parentally. For now, we can pass over those complications and consider the mathematical role that the fitness concept plays in the two models just described. Whether selection acts on viabilities or on fertilities, fitness describes how the frequencies of traits can be expected to change.

Even though I have divided the life cycle in two and thereby distinguished viability and fertility selection, it is perfectly possible for both sorts of selection to influence the evolution of a trait. This will occur when the trait affects both viability *and* fertility. When this happens, the two sorts of selection can come into conflict.

The term "sexual selection" is often applied to the evolution of traits that augment fertility but impair viability. Consider the peacock's tail. Having a big showy tail makes a peacock more attractive to females, so males with fancy tails tend to have more offspring than males without. On the other hand, a large showy tail makes peacocks more vulnerable to predators, so, from the point of view of viability alone, it would be better not to have one.

A model of this process would consider males with showy tails (S) and males with plain tails (P). S has lower viability than P, so in the part of the life cycle that goes from zygote to adult, S declines in frequency. However, S has greater fertility than P, so in the part of the life cycle that goes from the adult generation to the next generation of zygotes, S increases in frequency. What are the long-term consequences of this conflict? Everything depends on the magnitudes of the fitnesses. A showy tail represents good news (for fertility) and bad news (for viability); a plain tail is a trade-off of the opposite sort. The question is, Which trait is superior *overall?* Apparently, the reproductive advantages of a showy tail more than compensated for the cost paid in viability. This is a consequence of the specific biology of the organism considered. There is no *a priori* rule that says that fertility matters more than viability.

3.2 The Interpretation of Probability

Consider a language in which various propositions A, B, etc., are expressed. We can define a measure on those propositions, which we will call $P()$. This measure maps propositions onto numbers. $P()$ is a probability measure precisely when the following conditions are satisfied for any propositions A and B:

$0 \leq P(A) \leq 1$.
$P(A) = 1$, if A must be true.
If A and B are incompatible, then $P(A \text{ or } B) = P(A) + P(B)$.

These are Kolmogorov's (1933) axioms of probability.

We often apply the probability concept to examples in which we exploit background knowledge. For instance, in drawing from a standard deck of cards, we say that P(the card is a spade or a heart or a diamond or a club) = 1. We also say that P(the card is a spade or a heart) = P(the card is a spade) + P(the card is a heart). We might represent this fact about probabilities by talking about $P_M()$, which means that probabilities are assigned under the assumptions specified by a model M.

The mathematical concept of probability can be interpreted in different ways. I'll describe some of the main candidates that are available (Eells 1984) and then indicate how they bear on the problem of interpreting the probability concepts used in evolutionary theory.

The *actual frequency* of an event in a population of events is one possible interpretation that probability may be given. Suppose we toss a coin 100 times. On each toss, the coin lands either heads or tails. Let H be the proposition that the coin lands heads on some arbitrarily selected toss. $P(H)$ can be interpreted as the actual frequency of heads in the 100 tosses. Under this interpretation, all of the above axioms are satisfied. $P(H)$ is between 0 and 1 inclusive, and $P(H \text{ or } -H) = P(H) + P(-H) = 1$. The actual frequency interpretation of probability is an *objective* interpretation; it interprets probability in terms of how often an event actually happens in some population of events.

There is an alternative interpretation of probability that is *subjective* in character. We can talk about how much certainty or confidence we should have that a given proposition is true. Not only does this concept describe something psychological, it also is normative in its force. It describes what our *degree of belief ought to be*.

Again, let H be the proposition that a coin lands heads after it is tossed. The degree of belief we should have in this proposition must fall somewhere between 0 and 1. We can be maximally confident that the coin either will land heads or will fail to do so. And the degree of belief we assign to the coin's landing both heads and tails on a given toss should be 0. Degree of belief can be interpreted so that it satisfies the Kolmogorov axioms.

Many philosophers believe that science uses a notion of probability that is not captured by either the idea of actual relative frequency *or* by the subjective interpretation in terms of degrees of belief. We say that a fair coin has a probability of landing heads of 0.5 even when it is tossed an odd number of times (or not tossed at all). Similarly, a trait in a population can have a viability fitness of 0.5 even though its census size is not cut precisely in half in the passage from egg to adult. And we say that the probability that heterozygote parents will produce a heterozygote offspring is 0.5 even though we know that some such matings yield frequencies of heterozygotes that differ from 0.5. If we are describing an objective property of these systems, we are not talking about degrees of belief. Nor are we talking about actual frequencies. What could these probability statements mean?

A third interpretation of probability says that an event's probability is its *hypothetical relative frequency*. A fair coin need not produce exactly half heads and half tails when it is tossed a finite number of times. But if we were to toss the coin again and again, lengthening the number of tosses without limit, the frequency of heads would converge on 0.5. A probability value of x does not entail an *actual* frequency equal to x, but it does entail that the frequency in an ever-lengthening *hypothetical* sequence of tosses will converge on the value x.

Both the actual frequency and the degree of belief interpretations of probability say that we can define probability *in terms of something else*. If we are puzzled by what probability means, we can elucidate that concept by referring it to something else that, we hope, is less obscure. However, closer attention to the hypothetical relative frequency interpretation of probability shows that this interpretation offers no such clarification. For, if it is not overstated, this interpretation is actually circular.

To see why, consider the fact that an infinite series of tosses of a fair coin does not *have* to converge on a relative frequency of 0.5. Just as a fair coin can land heads up on each of the ten occasions on which it is tossed, so it can land heads up each time even if it is tossed forever. Of course, the probability of getting all heads in an infinite number of tosses is very small. Indeed, the probability of this happening approaches zero as the number of tosses is increased without limit. But the same is true for every specific sequence of results; for example, the probability of the alternation *HTHTHT*... approaches zero as the number of tries is increased. With an infinite number of tosses, each specific sequence has a zero probability of occurring; yet, one of them will actually occur. For this reason, we cannot equate a probability of zero with the idea of impossibility, nor a probability of one with the idea of necessity; this is why a fair coin won't *necessarily* converge on a relative frequency of 50 percent heads.

If the frequency of heads does not *have* to converge on the coin's true probability of landing heads, how are these two concepts related? The Law of Large Numbers (which I have just stated informally), provides the answer:

P(the coin lands heads / the coin is tossed) = 0.5
if and only if
P(the frequency of heads = $0.5 \pm e$ / the coin is tossed n times) approaches 1 as n goes to infinity.

Here, e is any small number you care to name. The probability of coming within e of 0.5 goes up as the number of tosses increases.

Notice that the probability concept appears on both sides of this if-and-only-if statement. The hypothetical relative frequency interpretation of probability is not really an interpretation at all, if an interpretation must offer a *noncircular* account of how probability statements should be understood (Skyrms 1980).

The last interpretation of probability I will discuss has enjoyed considerable popularity, even though it suffers from a similar defect. This is the *propensity interpretation of probability*. Propensities are probabilistic dispositions, so I'll begin by examining the idea of a dispositional property.

Dispositional properties are named by words that have "-ible" suffixes. Solubility, for example, is a disposition. It can be defined as follows:

> *X* is soluble if and only if, if *X* were immersed under normal conditions, then *X* would dissolve.

This definition says that an object is soluble precisely when a particular if/then statement is true of it. Notice that the definition allows for the possibility that a soluble substance may never actually dissolve; after all, it may never be immersed. Notice also that the definition mentions "normal conditions"; immersing a water-soluble object in water will not cause it to dissolve if the object is coated with wax. I've just described solubility as, so to speak, a "deterministic" disposition. According to the definition, soluble substances are not simply ones that *probably* dissolve when immersed in the right way—they are substances that *must* dissolve when immersed.

The propensity interpretation of probability offers a similar account of probabilistic if/then statements. Suppose the probability of a coin's landing heads, if it is tossed, is 0.5. If this statement is true, what makes it so? The suggestion is that probabilistic if/then statements are true because objects possess a special sort of dispositional property, called a *propensity*. If a sugar lump would dissolve when immersed, the sugar lump has the dispositional property of solubility; likewise, if a coin has a 0.5 probability of landing heads when tossed, the coin is said to have a propensity of a certain strength to land heads when tossed.

The propensity interpretation stresses an analogy between deterministic dispositions and probabilistic propensities. There are two ways to find out if an object is soluble. The most obvious way is to immerse it in water and see if it dissolves. But a second avenue of inquiry also is possible. Soluble substances are soluble because of their physical constitution. In principle, we could examine the physical structure of a sugar lump and find out that it is water soluble without ever having to dissolve it in water. Thus, a dispositional property has an *associated behavior* and a *physical basis*. We can discover whether an object has a given dispositional property by exploring either of these.

The same is true of probabilistic propensities. We can discover if a coin is "fair" in one of two ways. We can toss it some number of times and gain evidence that is relevant. Or, we can examine the coin's physical structure and find out if it is evenly balanced. In other words, the probabilistic propensities of an object can be investigated by attending to its behavior and also to its physical structure.

In spite of this apt analogy between probabilistic propensities and garden-variety dispositions, there still is room to doubt the adequacy of the propensity interpretation of probability. For one thing, the account is not general enough (Salmon 1984, p. 205, attributes this point to Paul Humphreys). When we talk about a soluble substance being disposed to dissolve when immersed, we mean that immersing it would *cause* it to dissolve. The if/then statement ("if it were immersed, then it would dissolve") describes a relation between cause and effect. However, there are many probability statements that do not describe any such causal relation. True, we can talk of the probability that an offspring will be heterozygote if its parents are heterozygotes. Here, the parental genotypes cause the genotype of the offspring. But we also can talk about the opposite relationship: the probability that an individual's parents were heterozygotes, given that the individual itself is a heterozygote. Offspring genotypes do not cause the genotypes of parents. Only sometimes does a conditional probability of the form $P(A/B)$ describe the causal tendency of B to produce A.

The more fundamental problem, however, is that "propensity" seems to be little more than a name for the probability concept we are trying to elucidate. In Molière's play *The Imaginary Invalid,* a quack announces that he can explain why opium puts people to sleep. The explanation, he says, is that opium possesses a particular property, which he calls a *virtus dormitiva* (a "dormitive virtue"). Molière's point was to poke fun at this empty remark. The quack has not explained why opium puts people to sleep since ascribing a dormitive virtue to opium is simply a restatement of the fact that taking opium will put you to sleep. I think a similar problem confronts the propensity interpretation of probability. We have no way to understand a coin's "propensity to land heads" unless we *already* know what it means to assign it a probability of landing that way. An *interpretation* of probability, to be worthy of the name, should explain the probability concept in terms that we can understand even if we do not already understand what probability is. The propensity interpretation fails to do this.

We now face something of a dilemma. The two coherent interpretations of probability mentioned so far are *actual relative frequency* and *subjective degree of belief*. If we think that probability concepts in science describe objective facts about nature that are not interpretable as actual frequencies, we seem to be in trouble. If we reject the actual frequency interpretation, what could it mean to say that a coin has an objective probability of landing heads of 0.5?

One possible solution to this dilemma is to deny that probabilities are objective. This is the idea that Darwin expresses in passing in the *Origin* (p. 131) when he explains what he means by saying that novel variants arise "by chance." "This," he says, "of course, is a wholly incorrect expression, but it serves to acknowledge plainly our ignorance of the cause of each particular variation." One might take the view that probability talk is always simply a way to describe our ignorance; it describes the degree of belief we have in the face of incomplete information. According to this idea, we talk about what *probably* will happen only because we do not have enough information to predict what certainly will occur.

Darwin could not have known that twentieth-century physics would block a thoroughgoing subjectivist interpretation of probability. According to quantum mechanics, chance is an objective feature of natural systems. Even if we knew everything relevant, we still could not predict with certainty the future behavior of the systems described in that physical theory.

If this were a text on the philosophy of physics, we could conclude that the subjective interpretation of probability is not adequate as an account of the probabilistic concepts deployed by the science in question. But the fact that chance is an objective matter in quantum mechanics tells us nothing about its meaning in evolutionary theory. Perhaps, as Darwin said, we should interpret the probabilistic concepts in evolutionary theory as expressions of ignorance and nothing else.

Before we embrace this subjective interpretation, however, another alternative should be placed on the table. Perhaps probability describes objective features of the world but cannot be defined noncircularly. This might be called an objectivist *no-theory theory of probability*. When we observe that a coin produces a certain actual frequency of heads in a run of tosses, we postulate that the coin has a given fixed probability of landing heads. This probability cannot be *defined* in terms of actual frequency or in any other noncircular way, but this does not mean that it is utterly unconnected with nonprobabilistic facts about the world. The Likelihood Principle (Section 2.2) describes how observed relative frequencies provide evidence for evaluating hypotheses about probabilities. And the Law of Large Numbers also helps us bring observations to bear on hypotheses about probabilities. By increasing sample size, we can increase our confidence that our probability estimate is correct (or is accurate to a certain specified degree).

The idea that probability can be defined noncircularly is no more plausible than the idea that a term in a scientific theory can be defined in purely observational language. An object's temperature is not correctly *defined* as whatever a thermometer says. Nor is intelligence correctly *defined* as whatever an IQ test measures. Both "definitions" ignore the fact that measuring devices can be inaccurate. A thermometer can provide *evidence* about temperature, and an IQ test can provide *evidence* about intelligence. Similarly, actual frequencies provide evidence about probabilities. Don't confuse the definition relation with the evidence relation; X can be evidence for Y even though X does not define what Y is.

A narrow empiricist (or an operationalist) would regard this relative autonomy of theory from observation as a defect in the theory. However, I would suggest that this relation of theory to observation should not bother us, provided that we still are able to test theories by appeal to observations. In similar fashion, a narrow empiricist will be disturbed by the fact that we sometimes use an objective concept of probability that cannot be defined in purely observational terms. The problem here is not with our use of probability but with the empiricist's scruples.

My argument so far has been that we should not reject the idea that probability is an irreducible and objective property simply on the grounds that it is irreducible. But this does not show that we should embrace such a concept. The subjectivist interpretation is still available, if not in quantum mechanics, then in most of the rest of science. When we describe an individual's fitness or a gene's chance of mutating, why do we assign numbers other than 1 or 0? The subjectivist will argue that our only reason is that we lack relevant information. If we know that the individual will die before reaching adulthood, its (viability) fitness is 0; if we know that it will reach adulthood, then its fitness is 1. We assign intermediate fitnesses, so the subjectivist says, because we do not know what will happen. Probability is merely a way for us to characterize our ignorance.

If we used probabilities only because we wish to make predictions, then the subjectivist would have a point. However, there is another reason to use probabilities. This pertains to the goal of capturing *significant generalizations*.

Consider the mating pairs in a population in which both parents are heterozygotes. These parental pairs produce different frequencies of heterozygote offspring. Although each obeys the usual Mendelian mechanism, the mating pairs differ from each other in various ways that account for their different frequencies of heterozygote offspring. We could describe these different mating pairs one at a time and list the unique constellation of causal influences at work in each. However, another strategy is to try to isolate what these parental pairs have in common. We do this when we describe each of them as participating in a Mendelian process in which P(offspring is *Aa* / parents are *Aa* and *Aa*) = 0.5.

It is important to recognize that this simple probability statement might be used to describe the parental pairs in the population *even if we possessed detailed information about the unique causal factors affecting each of them.* Our reason for using probability here is *not* that we are ignorant; we are not. We possess further information about the idiosyncratic details concerning each mating pair. These would be relevant to the task of prediction, but not necessarily to the task of explanatory description.

Levins (1966) proposes an analogy between biological models and maps. One of his points is that a good map will not depict every object in the mapped terrain. The welter of detail provided by a complete map (should such a thing be possible) would obscure whatever patterns we might wish to make salient. A good map depicts some objects *but not others.* (Of course, it is our interests that determine which objects are worth mapping.) In similar fashion, a good model of a biological process will not include every detail about every organism. In order to isolate general patterns, we abstract away from the idiosyncracies that distinguish some objects from others.

If we say that heterozygote parents have heterozygote offspring with a probability of 0.5, we are making a very general statement that goes beyond what we actually observe in some finite sample of heterozygote parents and their offspring. It isn't that our description of the sample is false. Rather, we assign a probability of 0.5 because we understand what we actually observe to be part of a much larger and more general class of systems. When we talk about "matings between heterozygotes," we have in mind a *kind* of event that may have many exemplifications. When we assign a probability to the different offspring that this type of mating may produce, we are trying to say something about what all exemplifications of that kind of event have in common. I see no reason why such statements cannot describe objective matters of fact.

Let us now leave the general question of whether probability can be viewed as an objective and irreducible property and consider the role of probability in evolutionary theory. I believe that the propensity interpretation of probability provides a useful account of the concept of fitness. Fitness *is* analogous to solubility. The only problem with the propensity interpretation is that it fails to provide a noncircular interpretation. To say that an organism's fitness is its propensity to survive and be reproductively successful is true but rather unilluminating.

I have already mentioned that the fitness (viability) of a trait need not precisely coincide with the actual frequency of individuals possessing the trait that survive to adulthood. Another way to see this point is to consider the fact that random genetic drift can cause changes in frequency when there is no variation in fitness. If two genes, *A* and *a*, have the same fitnesses, their frequencies may do a random walk. Given long enough, one or the other will go to fixation. If the fitness of a trait were defined as the actual frequency of individuals with the trait that survive, we would have to describe drift as a process in which genes differ in fitness. Evolutionists accept no such implication; they do not interpret the probabilistic concept of fitness in terms of *actual* frequencies.

In the previous section, I described a simple format for modeling viability and fertility selection. I pointed out that these models can be understood by using a simple rule of thumb: *Fitter traits increase in frequency, and less fit traits decline.* Now, in the light of the present discussion of what fitness means, I must qualify this rule of thumb. The models described earlier were ones in which we imagined that natural selection is the only cause affecting trait frequencies. Fitnesses determine the population's trajectory in this idealized circumstance but not otherwise.

How often is natural selection the only factor at work in a population? This question has a simple answer: *never*. Populations always are finite in size, which means that a trait's fitness plus its initial frequency do not absolutely determine its frequency after selection. The Law of Large Numbers is relevant here: The larger the population, the more probable it is that a trait with p as its frequency before selection and w_1 as its viability will have $pw_1/\overline{w} \pm e$ (for any small value of e) as its frequency after selection.

Population geneticists often say that models representing how natural selection works when no other evolutionary forces are present assume "infinite population size." This idealization allows us to be certain of (i.e., assign a probability of 1 to) the predictions we make about trait evolution based just on trait fitnesses and frequencies.

If populations never are infinitely large, what is the point of considering such obviously false models? The point is that if populations are large (though finite), one can be "almost certain" that the predictions of the model are correct. A fair coin that is tossed ten times has a good chance of not producing between 4 and 6 heads, but it has almost no chance at all of falling outside of the 40,000 to 60,000 range for heads when it is tossed 100,000 times. With large sample size, the predictions calculated for an infinite population are plenty close enough.

Every model involves simplifications. Many evolutionary forces impinge simultaneously on a population. The evolutionist selects some of these to include in a mathematical representation. Others are ignored. The model allows one to predict what will happen or to assign probabilities to different possible outcomes. All such models implicitly have a *ceteris paribus* clause appended to them. This clause does not mean that all factors not treated in the model have *equal* importance but that they have *zero* importance. The Latin expression would be more apt if it were *ceteris absentibus* (Joseph 1980). Models can be useful even when they are incomplete if the factors they ignore have small effects. This means that an evolutionary model is not defective just because it leaves out something. Rather, the relevant question is whether a factor that was ignored in the model would *substantially* change the predictions of the model if it were taken into account.

3.3 Two Ways to Find Out About Fitness

As the propensity interpretation of fitness (Mills and Beatty 1979) states, there are two ways to find out about the fitnesses of traits in a population. Although a trait's fitness is not defined by its actual degree of survivorship and reproductive success, looking at these actual frequencies provides evidence about the fitness of the traits. If the individuals with trait *A* survive to adulthood more often than the individuals with trait *B*, this is evidence that *A* is fitter than *B*. The inference from actual frequencies to fitnesses is mediated by the Likelihood Principle (Section 2.2). If we observe that *A* individuals outsurvive *B* individuals, this observation is made more probable by the supposition that *A* is fitter than *B* than it is by the supposition that *B* is fitter than *A* or by the supposition that the fitnesses are equal.

There is another way to find out about fitness besides observing actual frequencies. Recall that we can find out if an object is soluble without having to immerse it in water—we can examine its physical makeup. If we possess a theory that tells us which physical properties make an object water soluble, we can keep the object dry and still say whether it is soluble. In similar fashion, we can reach judgments about an organism's fitness by examining its physical makeup.

This second approach to fitness can issue from scientific common sense and also from sophisticated theorizing. When we note that zebras are hunted by predators, it becomes plausible to think that faster running speed makes for a fitter zebra. No fancy mathematical model is needed to see the point of this idea. We think of the zebra as a machine and ask how an engineer might equip it for better survival and reproductive success. Although hunches about fitness that derive from such thinking may be mistaken, it seems undeniable that such considerations can generate plausible guesses about which traits will be fitter than which others.

The same thought process occurs at a more sophisticated level when we use mathematical models. Once we understand Fisher's sex ratio argument (Section 1.4), we see that in certain sorts of populations, a fitness advantage goes to a parent who produces offspring solely of the minority sex. We base this judgment on a model, not on the empirical observation of how many grandoffspring various parents happen to have.

In Section 1.6, I mentioned that each cause of evolution can be understood both in terms of its *consequences* and in terms of its *sources*. Natural selection occurs when there is variation in fitness. This variation may have the consequence that some traits increase in frequency while others decline. In addition, the variation in fitness that occurs within a population will have its source in the complex nexus of relationships that connects organisms to their environments and to each other. Because the fitnesses of traits have their sources as well as their consequences, we can find out about fitness in the two ways just described.

Ideally, we can pursue both modes of investigation simultaneously. The population exhibits variation, and so we are able to measure differences in viability and fertility. We also can find out what it is in the environment that induces these fitness differences, perhaps by experimentally manipulating the organism/environment relationship. Kettlewell's (1973) study of industrial melanism in the peppered moth *Biston betularia,* for example, involved both lines of inquiry. Kettlewell tried to measure whether dark moths have higher mortality rates than light moths. In addition, he manipulated the environment to find out whether dark moths sitting on dark trees are less vulnerable to predation than light moths sitting on the same trees; symmetrically, he also investigated whether light moths on trees not darkened by pollution avoid predators more successfully than dark moths on the same trees. This dual line of investigation led to two conclusions: (1) Dark moths are fitter than light ones in polluted areas, but the reverse is true in unpolluted areas, and (2) these fitness differences are due to the fact that moths that match the trees on which they perch are less visible to predators. My point is not that Kettlewell's investigation was flawless but that he tried to get at fitness differences by looking both at consequences *and* at sources.

This is the ideal case; in practice, evolutionists often face problems that cannot be treated in this way. For example, suppose one is studying a trait that is universal in the population of interest. One may suspect that it evolved because natural selection favored it over the alternatives that were present in some ancestral population. The problem is that the other variants that were present ancestrally are no longer around. The big brain found in human beings may have a selective explanation, but what were the specific alternatives against which it competed? What was the environment like in which the competition took place? It isn't that these questions are unanswerable but that they may be difficult to answer. Kettlewell had it easy, we might say, because the variation was in place and the environment he needed to consider was the one he could actually observe. Investigators who reach deeper into the past typically are not so lucky.

The problem just described arises from the fact that selection tends to destroy the variation on which it acts (this statement will be made more precise in Section 5.2). This raises an epistemological difficulty since we must know about ancestral variation if we are to reconstruct the history of a selection process. Selection tends to cover its own tracks, so to speak.

I have emphasized the difficulty of figuring out what the variants were against which a given trait competed. But even if the field of variation is plain to see, it still can be hard to determine what the sources of fitness differences are. Dobzhansky repeatedly discovered fitness differences among various chromosome inversions in *Drosophila*. The phenotypic consequences of these inversions were difficult to identify, and so it often was quite unclear *why* one inversion was fitter than another. Traits do not always wear their adaptive significance on their sleeves.

3.4 The Tautology Problem

Herbert Spencer described Darwin's theory with the phrase "the survival of the fittest." Ever since, this little slogan has been used by various people to challenge the scientific status of the theory of evolution: Who survives? Those who are the fittest. And who are the fittest? Those who survive. The idea is that evolutionary theory is untestable because fitness is *defined* in terms of actual survivorship. Given this definition, it cannot fail to be true that the organisms we presently observe survived because they were the fittest. The theory is said to be a "tautology" and therefore not an empirical claim at all. Creationists (e.g., Morris 1974) have pressed this charge, but so have others. The criticism is persistent enough that it is worth seeing why it is misguided.

Before I address the criticism, the term "tautology" needs to be clarified. The first important point is that *propositions* are the only things that are tautologies. Not all propositions are tautologies, but all tautologies are propositions. A proposition is what is expressed by a declarative sentence in some language; it is either true or false. But notice that the phrase "the survival of the fittest" is not a declarative sentence. If we are going to assess whether "the survival of the fittest" is a tautology, we first must be precise about which proposition we wish to examine.

What makes a proposition a tautology? Logicians apply this term to a special class of simple logical truths. "It is raining or it is not raining" is a tautology because it has the form *P or not-P*. The definitions of the logical terms "or" and "not" suffice to guarantee that the proposition is true; we don't have to attend to the nonlogical vocabulary in the sentence (e.g., "raining"). The sentence "it is raining or it is not raining" is true for the very same reason that "pigs exist or pigs don't exist" is true. This has nothing to do with rain or with pigs.

The term "tautology" is sometimes given a wider application. Consider the sentence "for all *x*, if *x* is a bachelor, then *x* is unmarried" (or, more colloquially, "all bachelors are unmarried"). The meaning of the logical terms in this sentence do *not* suffice to guarantee that it is true. The logical terms are "all" and "if/then." Their meanings are not enough; in addition, you need to know the meanings of the nonlogical terms ("bachelors" and "unmarried"). The truth of the quoted sentence follows from the definitions of the terms it contains. Philosophers label such sentences *analytic*. Statements whose truth or falsity is not settled by the meanings of the terms they contain are called *synthetic*.

The charge that "survival of the fittest" is a "tautology" might be formulated, then, as the claim that some proposition is analytic. But which proposition are we talking about? Perhaps the following is the proposition to consider:

> The traits found in contemporary populations are present because those populations were descended from ancestral populations in which those traits were the fittest of the variants available.

Notice, first of all, that this statement is not a tautology; it is not a truth of logic that present populations were descended from ancestral populations. This implication of the statement is true enough, but it is no tautology. The second thing to notice is that the statement, taken as a whole, is false. A trait now at fixation in some population may have reached fixation for any number of reasons. Natural selection is one possible cause, but so are random genetic drift, mutation, and migration.

Incidentally, it is a curiosity of some creationist argumentation that evolutionary theory is described as being (1) untestable, (2) empirically disconfirmed, and (3) a tautology. This nested confusion to one side, the main point here is that the statement displayed above is not a tautology and, in any case, is not part of the theory of evolution. Far from being an analytic truth, it is a synthetic falsehood.

In saying that the statement is not part of evolutionary theory, I am not saying that the theory contains no tautologies. Perhaps the following is a serviceable definition of *fitness*:

> Trait X is fitter than trait Y if and only if X has a higher probability of survival and/or a greater expectation of reproductive success than Y.

There is room to quibble with the adequacy of this statement, but fine points aside, it is a reasonably good definition of fitness.

The fact that the theory of evolution *contains* this tautology does not show that the whole theory *is* a tautology. Don't confuse the part with the whole. Perhaps what is most preposterous about the "tautology problem" is that it has assumed that the status of the whole theory depends on the verdict one reaches about one little proposition (Kitcher 1982a; Sober 1984b).

The two main propositions in Darwin's theory of evolution are both *historical hypotheses* (Section 1.4). The ideas that all life is related and that natural selection is the principal cause of life's diversity are claims about a particular object (terrestrial life) and about how it came to exhibit its present characteris-

Box 3.1 Quine on *A Priori* Truth

How can we tell if a statement that strikes us as obvious is a definitional truth? We may look at the statement and think that no observation could possibly count against it. But perhaps this simply reflects our lack of imagination, not the fact that the statement really is *a priori* (i.e., justifiable prior to or independent of experience).

Duhem's Thesis should lead us to take this problem seriously. Consider what it means to say that *H* is not an empirical claim, if we accept the idea that hypotheses have empirical consequences only when auxiliary assumptions are conjoined with them. We are saying that it never will be reasonable for us to accept auxiliary assumptions *A* that could be added to *H* such that *H&A* make predictions that do not follow from *A* alone. This requires a kind of omniscience about the future of science, one that the history of science has taught us we do not possess.

Immanuel Kant thought that Euclidean geometry and the thesis of determinism are *a priori* true. He believed that no observation could count against either. But in our century, it was discovered that Euclidean geometry, when conjoined with an independently plausible physical theory, makes predictions that turn out to be false. And determinism likewise yields false predictions when it is conjoined with a plausible background theory. The former insight derives from relativity theory, the latter from quantum mechanics. Kant did not foresee these new theoretical developments.

Based partly on examples such as these, Quine (1952, 1960) concluded that there are no *a priori* (or analytic) truths. I will not discuss this radical conclusion here. The more modest point is that we should be circumspect when we say that this or that proposition is a tautology. How can we tell that what seems to be a definitional truth really is one?

tics. It is quite clear that neither of these hypotheses can be deduced from definitions alone. Neither is analytic. Darwin's two-part theory is no tautology.

Let's shift our attention to another class of statements in evolutionary theory and consider the general if/then statements that models of evolutionary processes provide. Are these statements empirical, or are they definitional truths? In physics, general laws such as Newton's Law of Gravitation and the special theory of relativity are empirical. In contrast, many of the general laws in evolutionary biology (the if/then statements provided by mathematical models) seem to be nonempirical. That is, *once an evolutionary model is stated carefully, it often turns out to be a (nonempirical) mathematical truth.* I argued this point with respect to Fisher's sex ratio argument in Section 1.5.

Now let's consider another example. The Hardy-Weinberg Law is sometimes given the following rough formulation:

> If there are p *A* genes and q *a* genes at some locus in a population, then the frequencies of the three genotypes *AA*, *Aa*, and *aa* will be p^2, $2pq$, and q^2, respectively.

The idea is that one can compute the frequencies of different genotypes in organisms from the frequencies of the gametes that produce them. The following table is usually given as an explanation of why the Hardy-Weinberg Law is true.

			Mother	
			p	q
			A	a
Father	p	A	p^2	pq
	q	a	pq	q^2

The indented statement displayed here requires qualification. We need to assume random mating and that the frequencies of the alleles in the two sexes are the same. In addition, we need to assume that the gamete frequencies are taken right before fertilization and that the offspring are censused immediately after fertilization (so there is no time for selection to throw in a monkey wrench). With infinite population size, the genotype frequencies follow from the gametic frequencies.

Given all these provisos, the Hardy-Weinberg Law seems to have the same status as the following proposition about coin tossing:

> If two coins are tossed independently, where each has a probability p of landing heads and q of landing tails, then the probabilities of getting two heads, one head and one tail, and two tails are p^2, $2pq$, and q^2, respectively.

This proposition about coin tossing is a mathematical truth; it is a consequence of the mathematical terms it contains. The statement follows from the probability axioms and from the definition of "probabilistic independence." The same holds true for the Hardy-Weinberg Law. Observations are quite unnecessary to verify either proposition.

If we use the term "tautology" sufficiently loosely (so that it encompasses mathematical truths), then many of the generalizations in evolutionary theory are tautologies. What is more, we seem to have found a difference between physics and biology. Physical laws are often empirical, but general models in evolutionary theory typically are not.

For the logical positivists, physics was the paradigm science; for them, physics mainly meant Newtonian mechanics, relativity theory, and quantum mechanics. Indeed, these three bodies of theory *do* contain empirical laws. It was only a short jump from these examples to the general thesis that a scientific theory *is* a set of empirical laws.

This view of science puts evolutionary theory in double jeopardy. First, theories cannot be historical hypotheses, so it is a misnomer to talk about Darwin's "theory" of evolution. Second, the truly general parts of evolutionary theory often are not empirical.

The word "tautology" has a pejorative connotation. It doesn't just mean a mathematical truth but an empty truism. This negative implication lies behind the claim that evolutionary theory is a "mere" tautology. Yet, no one seems to dismiss the work of mathematicians as "mere tautologies." The reason is that mathematics can be deep and its results are nonobvious.

The same point applies to a great deal of model building in evolutionary theory. Perhaps Fisher's sex ratio argument, construed as an if/then statement, is a mathematical truth. Even so, it is very far from being trivial. And it was not obvious until Fisher stated the argument. Thanks to his insights, we now may be able to find obvious what earlier had been quite unclear.

Physics worship and a mistaken picture of mathematics as a trivial enterprise might lead one to dismiss model building in evolutionary biology as not genuinely "scientific"—the models are not empirical but are "mere" mathematics. However, why be seduced by this double error? "Science" should be used as a term that encompasses *all* the sciences. If there is more than one kind of science—if the sciences differ from each other in interesting ways—we need to acknowledge this fact and understand it. There is no point in withholding the label of "science" from evolutionary biology just because it isn't exactly like physics. Of course, the theory contains "tautologies" (mathematical truths); every theory does. Some of these "tautologies" are interesting and important guides to our understanding of the living world. And there is more to the science than its general mathematical models. Historical hypotheses describe properties of the particular objects found in the tree of life. These hypotheses are empirical.

3.5 Supervenience

The physical properties of an organism and the environment it inhabits determine how fit that organism is. But the fitness that an organism possesses—how viable or fertile it is—does not determine what its physical properties must be like. This asymmetrical relation between the physical properties of the organism in its environment and the fitness of the organism in its environment means that fitness *supervenes* upon physical properties (Rosenberg 1978, 1985).

Here is another way to formulate the supervenience thesis: If two organisms are identical in their physical properties and live in physically identical environments, then they must have the same fitness. But the fact that two organisms have the same probability of survival or the same expected number of offspring does not entail that they and their environments must be physically identical. A cockroach and a zebra differ in numerous ways, but both may happen to have a 0.83 probability of surviving to adulthood.

The idea of supervenience can be defined more generally. One set of properties P supervenes on another set of properties Q precisely when the Q properties of an object determine what its P properties are—but not conversely. If P supervenes on Q, then there is a one-many mapping from P to Q (Kim 1978).

The fact that fitness supervenes on physical properties suggests a more general thesis: *All biological properties supervene on physical properties.* And this thesis about the properties investigated in biology suggests a more general thesis still: *All the properties investigated in sciences other than physics supervene on physical properties*. This supervenience thesis assigns to physics a special status among all the sciences. It asserts that the vocabulary of physical properties provides the most fine-grained description of the particular objects that populate the world.

In Section 1.6, I presented a thesis that I labeled *physicalism,* which says that every object is a physical object. According to physicalism, psychology and biology have as their domains the physical objects that have minds or are alive. Physics seeks to characterize what all physical objects have in common; its domain includes the domains of the other sciences.

The concept of supervenience is a useful tool for making physicalism more precise. To see why, we must consider what it means for an object to be "physical." It doesn't simply mean that *some* of the object's properties are physical; after all, if an organism had an immaterial soul or an immaterial *élan vital,* it could still have a mass and a temperature. Nor does it mean that *all* of the object's properties are physical. Consistent with physicalism, an organism may have a particular fitness value and a love of music, even though these properties are not discussed in physics.

What, then, could the physicalist mean by saying that all objects are "physical objects?" The suggestion is this: *To say that an organism is a physical thing is to say that all its properties supervene on its physical properties*. The concept of supervenience provides a more precise rendition of what physicalism asserts.

One question raised in Section 1.6 about the thesis of physicalism was whether physics, in principle, is capable of explaining everything. The supervenience thesis now before us allows us to pursue that question further.

Suppose we notice that two chromosome inversions change frequency in a population of *Drosophila* in the course of a year. Investigation reveals that the changes are due to selection. We discover that one type has a higher viability than the other, and so we explain the change in frequency by saying that the one type had a greater fitness value than the other.

We then inquire as to the physical basis of this difference in fitness. We discover that the one chromosome inversion produces a thicker thorax, which insulates the fly better against the low temperatures that prevail. Once this physical characterization is obtained, we no longer need to use the word "fitness" to explain why the traits changed frequency. The fitness concept provided our initial explanation, but the physical details provide a deeper one. This does not mean that the first account was entirely unexplanatory. Fitness is not the empty idea of a dormitive virtue. The point is that although fitness is explanatory, it seems to be a placeholder for a deeper account that dispenses with the concept of fitness. Instead of saying that one chromosome inversion had a higher fitness than the other, we can say that the first one produced a thicker thorax than the other and that this difference explains why the first type of fly outsurvived the second.

This example suggests a general claim about the relationship of fitness to the physical properties of an organism. Suppose that F is the set of fitnesses that characterize the organisms in some population and $M(F)$ is the set of physical properties of those organisms on which F supervenes. The claim we now can consider is that $M(F)$ explains whatever F explains. This thesis can be generalized. Let *B* be the biological properties that characterize the organisms in a population and *M*(*B*) be the physical properties on which those biological properties supervene. We now can consider the thesis that *M*(*B*) explains whatever *B* explains. The still more general thesis that this suggests is that *there is a physical explanation for any phenomenon explained by sciences outside of physics.*

Is this thesis about explanation rendered plausible by the supervenience thesis? If biological properties supervene on physical properties, does it follow that physical properties can explain whatever biological properties explain? Putnam (1975) has argued that the answer is no. Consider his very simple example. Suppose we have a board with a round hole in it that is 6 inches in diameter. We try to pass a square peg, which is 6 inches on each side, through the hole, but we fail. How are we to explain the fact that the peg did not pass through the hole? Putnam said that the size and shape of the hole and the peg provide the obvious explanation. Call these the *macroproperties* of the system. Alternatively, we could characterize the position and other properties of each of the atoms in the peg and the board. Do these *microproperties* explain why the peg would not go through the hole?

The macroproperties supervene on the microproperties. The positions of each of the atoms in the board and peg determine the macro shapes and sizes, but the converse is not true. If the macroproperties explain why the peg would not go through the hole, do the microproperties also explain this fact? Putnam says they do not. The exhaustive list of microproperties presents a great deal of irrelevant information. The exact position of each atom does not matter: The microstory is not explanatory, according to Putnam, because it cites facts that are inessential.

Putnam's proposal has the quite general consequence that if *X* properties supervene on *Y* properties, then *Y* properties never explain what *X* properties explain. If fitness explains the change in trait frequency that occurs in the *Drosophila* population, then the fly's thick thorax does not. Surely this conclusion has a peculiar ring to it.

Putnam's argument relies on the following assumption: If *C* is not necessary for the occurrence of *E*, then *C* is not relevant to explaining *E*. Putnam says that the positions of the atoms are explanatorily irrelevant on the grounds that the peg would have failed to pass through the hole even if the atoms had been arranged somewhat differently. But surely this is a mistaken constraint to place on the concept of explanation. The explanation for Moriarty's death is that Holmes shot him. True, Holmes could have used another weapon, and Moriarty would have died if someone else had done the deed. Holmes's firing the gun was not *necessary* for Moriarty's death, yet, it is a perfectly satisfactory explanation of Moriarty's death to say that Holmes pulled the trigger.

What does seem plausible is that the enormously long list of individual atoms and their properties is a needlessly verbose explanation for why the peg failed to pass through the hole. An explanation crowded with boring details is still an explanation, though it is an inferior one. Perhaps, then, we should reformulate Putnam's thesis; the idea is not that microaccounts are unexplanatory but that they are poor explanations.

The trouble with this suggestion is that microstories often *do* provide enhanced illumination. For example, my headache disappeared because I took aspirin. It is not a useless exercise to describe the way the molecules in the aspirin acted on the various subparts of my body. Surely microaccounts sometimes do a very good job of explaining what macroaccounts explain.

Physics has a domain that subsumes the domains of psychology and biology. In addition, it is at least arguable that every event that has a biological or psychological explanation also has a physical explanation. Of course, it often will be impossible for us to *state* that physical explanation, perhaps because of our ignorance or because writing out the physical explanation would take too much ink.

This tentative conclusion makes biology appear to have no explanatory problems it can call its own. *In principle,* physics is able to explain any event that biology can hope to explain, though *in practice,* our limited knowledge may prevent us from stating the physical explanation. The autonomy of biology therefore seems to depend on our ignorance of the world. The reason we have separate sciences is not that there are different kinds of explanatory problems to be addressed. Rather, the division of labor among the sciences is simply a convenient strategy: We find it easier to attack different problems by using different vocabularies.

Although I do not think that this conclusion is an affront to the dignity of biology as a discipline, there is another way to think about the relationship of biology to physics. Thus far, I have focused on the problem of explaining *single events*. Trait frequencies change in a *Drosophila* population, and we want to know why. However, there is another goal in science besides the explanation of single events. The different sciences also seek to construct descriptive frameworks that characterize what various single events have in common. In addition to explaining single events, we also want to describe *general patterns*.

It is here that the vocabulary of supervening properties makes an irreducible contribution. Consider the example with which I began this section—the concept of fitness. Models of natural selection describe how a population changes in response to the variation in fitness it contains. For example, Fisher's (1930) so-called *fundamental theorem of natural selection* says that the rate of evolution in a population (when natural selection is the only force at work) is equal to the (additive genetic) variance in fitness. This generalization subsumes evolution in orchids, iguanas, and people. These different populations differ from each other in countless physical ways. If we were to describe only their physical characteristics, we would have to tell a different story about the evolution of each of them. But by abstracting away from these physical differences, we can see that there is something these different populations have in common. Fisher's generalization about natural selection cannot be reduced to physical facts about living things precisely because fitness supervenes on those physical facts.

The same can be said about many other generalizations in biology. For example, the Lotka-Volterra equations in ecology describe how the number of predators and the number of prey organisms are dynamically related. These equations apply to any pair of populations in which organisms in one prey upon organisms in the other. What is this relation of *predation*? Lions prey on antelopes; Venus's-flytraps prey on flies. What does a lion have in common with a flytrap that makes both of them predators? It isn't in virtue of any physical similarity that these two organisms both count as predators. True, lions *catch and eat* antelopes, and flytraps *catch and eat* flies. But the physical details of what catching and eating mean in these two cases differ markedly. Biological categories allow us to recognize similarities between physically distinct systems.

So what answer can we give to the question of whether physics can explain everything that biology can explain? First, we need to divide the question in two: (1) If there is a biological explanation for *why some particular event occurs*, is there also a physical explanation? (2) If there is a biological explanation of *what several particular events have in common*, is there also a physical explanation? Perhaps the answer to (1) is yes; as for (2), the answer I would give is no. It may be that each single event has a physical explanation, but this does not mean that every pattern among events can be characterized in the vocabulary of physics (Sober 1984b).

3.6 Advantageousness and Fitness

We often use the terms "fitness" and "advantage" interchangeably. We say that it is advantageous for zebras to run fast (rather than slowly) when attacked by predators. We also say that fast zebras are fitter than slow zebras. Although there is a special circumstance in which these two descriptions are equivalent, in general they are not.

To see why, let us consider a population in which selection acts on two characteristics at once. Suppose that a zebra population experiences selection on running speed *and* selection on disease resistance. For simplicity, imagine that the organisms in the population are either *Fast* or *Slow* and that they are either *Resistant* to the disease or *Vulnerable* to it.

In principle, there are four combinations of traits that an organism might possess, which are displayed in the following 2 x 2 table. The entries represent the fitnesses of each combination. The absolute values don't matter; just attend to the inequalities they imply.

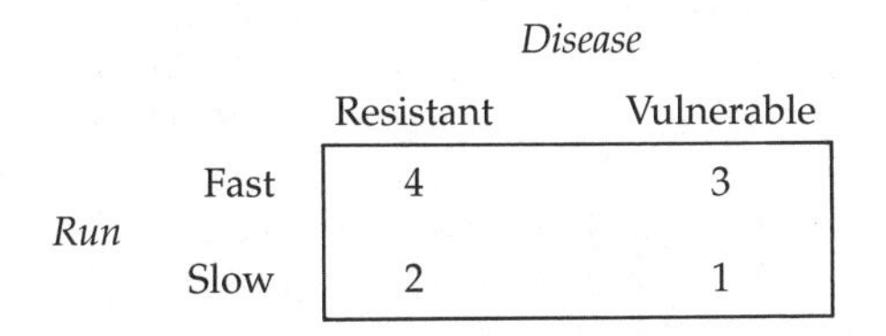

		Disease	
		Resistant	Vulnerable
Run	Fast	4	3
	Slow	2	1

The best combination of traits is to run fast and be disease resistant; the worst is to run slow and be vulnerable to disease. How should we rank the two intermediate combinations? Suppose the disease in question is sufficiently rare

Box 3.2 Reduction

Scientists sometimes talk about "reductionism" and about "reducing" one theory (or process or phenomenon) to another. Philosophers have written a great deal about how these ideas should be understood (reviewed in Wimsatt 1979).

One reading of what reduction means is suggested by the accompanying text. Perhaps "*X* reduces to *Y*" means that *Y* can explain whatever *X* can explain but not conversely. There is a simple objection to this suggestion. Presumably, *X*&*Z* can explain whatever *X* explains, but the converse is not true. However, it surely trivializes the concept of reduction to say that *X* reduces to the conjunction *X*&*Z*, where *X* and *Z* are quite unrelated theories.

A related point of departure has been the idea that *reduction* means *deduction*. To reduce theory *X* to theory *Y* is to deduce *X* from *Y*. The first complication arises when we recognize that the two theories may contain different vocabularies. In this case, the deduction requires that the reducing theory *Y* be supplemented with bridge principles *B* that show how the two vocabularies are connected. The proposal is that *X* reduces to *Y* when *X* can be deduced from *Y*&*B*.

One problem this proposal faces is that scientists often talk about reducing one theory to another even though the reduced theory is, at best, an approximation. For example, Mendel's Law of Independent Assortment is often interpreted as saying that any two genotypes are statistically independent of each other. This general statement isn't true if the relevant genes are on the same pair of chromosomes. If Mendel's law is false, it cannot be deduced from true propositions of any sort. Yet, we talk about reducing Mendel's theory to the chromosome theory of inheritance. As a result, it has been suggested that in reduction we deduce a "corrected" version of the reduced theory (Schaffner 1976). The problem is to spell out what "corrected" means. What is the difference between reducing one theory *to* another and refuting one theory *by* another (Hull 1976)?

"Reductionism" is used in a quite different sense when it is applied to research programs. Suppose one research program assumes that a given phenomenon is influenced by causal factors $C_1, C_2, \ldots, C_n$. A new research program is then announced that tries to show that some of those *n* variables are dispensable. The new program will undoubtedly be called *reductionistic*. Rather than postulating *n* causes, it aims to establish that the number of relevant independent variables can be reduced. *Adaptationism* (Chapter 5) and *sociobiology* (Chapter 7) have been termed reductionistic in this sense.

and that predators are sufficiently common that it would be better to be fast and vulnerable to the disease than slow and resistant to it. This is the fitness ordering represented in the table.

Notice that running fast is *advantageous*. Regardless of whether you are resistant to the disease or vulnerable to it, you do better by running fast than by running slow (4 > 2 and 3 > 1). The same reasoning implies that being disease resistant is *advantageous*. Regardless of whether you run fast or slow, you do better by being resistant to the disease than by being vulnerable to it (4 > 3 and 2 > 1).

What will happen in the population if natural selection acts simultaneously on the variation for running speed and the variation for disease resistance? If we use the rule that says that fitter traits increase in frequency and less fit traits decline, can we conclude that the fittest of the four combinations will evolve to fixation?

This will not be true if there is a strong correlation between running fast and being vulnerable to disease. For simplicity, imagine that every organism in the population is in either the upper right or lower left box of the 2 x 2 table. Natural selection now has *two* trait combinations to act upon, not *four*. If so, *Fast&Vulnerable* will evolve and *Slow&Resistant* will exit from the scene. The upshot is that a *dis*advantageous trait (being vulnerable to disease) increases in frequency and an advantageous trait (being resistant to disease) declines.

There is nothing wrong with the rule that says that fitter traits increase in frequency and less fit traits decline. (Of course, you need to remember that using the rule requires the assumption that natural selection is the only force at work and that the traits are heritable.) The important point is that the rule says that *fitter* traits evolve, not that *advantageous* traits always do so.

What is the fitness of the trait of being fast? It is an average. Individuals who run fast may be disease resistant (in which case they have a fitness of 4) or they may be vulnerable to disease (in which case they have a fitness of 3). So the fitness of *Fast* is a weighted average that falls somewhere between 4 and 3, the weights reflecting how often the trait occurs in these two contexts. Likewise, the fitness of *Slow* must fall somewhere between 2 and 1. From this, it follows that *Fast* must be fitter than *Slow*.

The same is *not* true of *Resistant* and *Vulnerable*. The former trait's fitness must fall between 4 and 2, the latter's between 3 and 1. Which trait actually has the higher fitness depends on whether this pair of traits is correlated with other traits that have an impact of their own.

If the two traits evolve *independently*, then advantageous traits will have a higher fitness. I already have shown that *Fast* is fitter than *Slow*, regardless of how much correlation there is between running speed and disease resistance. Now let's consider what the fitnesses of disease resistance and disease vulnerability are if resistance is independent of running speed. Suppose there are p *Fast* individuals in the population and q *Slow* ones ($p + q = 1$). If running speed is independent of disease resistance, then the fitness of *Resistant* is $4p + 2q$ while the fitness of *Vulnerability* is $3p + 1q$. The advantageous trait is fitter, if the independence assumption holds true.

What could cause correlation of characters? One answer is *pleiotropy*, which occurs when a single gene has two phenotypic effects. If the A allele causes phenotypes P_1 and Q_1 and the a allele causes phenotypes P_2 and Q_2 in a population of haploid organisms, then P_1 and Q_1 will be correlated. The phenotypic correlation is due to the fact that the two phenotypes have a genetic common cause.

Rose and Charlesworth (1981) describe an interesting example of "antagonistic pleiotropy" in female *Drosophila*. Females with high fecundity early in life tend to lay fewer eggs when they are older. There is a correlation between high fecundity at one developmental stage and low fecundity at another. The fittest conceivable fly, so to speak, will have high fecundity at both stages. But

Box 3.3 Correlation

Consider two dichotomous (on/off) traits. For example, suppose the people in a population either smoke (S) or do not and that they either get lung cancer (C) or do not. Smoking is positively correlated with lung cancer precisely when

$$P(C/S) > P(C/-S).$$

Positive correlation means that the frequency of cancer among smokers exceeds the frequency of cancer among nonsmokers. For negative correlation, reverse the inequality sign. For zero correlation (independence), replace the inequality with equality.

Positive correlation does not require that most smokers get cancer. The previous inequality should not be confused with

$$P(C/S) > P(-C/S).$$

If 10 percent of the smokers get cancer but only 2 percent of the nonsmokers do, smoking and cancer are positively correlated.

Correlation is a symmetrical relation; if smoking is correlated with cancer, then cancer is correlated with smoking. The inequality that defines positive correlation can also be written as follows:

$$P(S/C) > P(S/-C).$$

A consequence of this symmetry is that correlation and causation must be different. Causation is not symmetrical; the fact that smoking causes cancer does not entail that cancer causes smoking.

It is possible for two traits to be correlated even though neither causes the other. This can happen when they are joint effects of a common cause. A drop in the barometer reading today is correlated with a storm tomorrow, but neither causes the other. Each is an effect of today's weather conditions.

It also is true that cause and effect do not have to be correlated. Suppose smoking promotes heart attacks but that smoking is correlated with some other factor that tends to prevent heart attacks. For example, suppose that smokers tend to eat low cholesterol diets and that nonsmokers tend to eat foods high in cholesterol. If smoking causally promotes heart attacks to the same degree that low cholesterol tends to prevent heart attacks, it may turn out that $P(H/S) = P(H/-S)$. Indeed, if low cholesterol prevents heart attacks more powerfully than smoking promotes them, it may turn out that $P(H/S) < P(H/-S)$. Thus, a causal factor and its effect may be positively correlated, uncorrelated, or negatively correlated.

To say that a trait is evolutionarily advantageous is to say that it *causally promotes* survival and/or reproductive success. To say that a trait is fitter than its alternative is to say that it is *correlated* with survival and/or reproductive success. Because cause and correlation are different, there is a difference between saying that a trait is advantageous and saying that it is fitter than its alternative.

because of a genetically induced correlation, this combination of characters is not available for selection to act upon.

A second mechanism that can produce correlation of characters is genetic *linkage*. Again imagine a haploid organism in which the *A*-locus and the *B*-locus are close together on the same chromosome. At each locus, there are two alleles. The linkage between the loci means that the independence assumption fails. With perfect correlation of the alleles (*A,a,B,b*), the population contains only two combinations (*AB* and *ab*), rather than four. If each allele has its own phenotypic effect, the result will be a correlation of phenotypic characters.

The term "supergene" is sometimes applied to a set of strongly linked genes that contribute to the same or to related phenotypes. In many plants, outcrossing is promoted by a mechanism called "heterostyly." The plants come in two forms: *Thrums* have short styles and tall anthers, and *pins* have the reverse arrangement. In the primrose (*Primula vulgaris*), there is linkage between the gene for short style and the gene for tall anthers (Ford 1971). The correlation induced by linkage in this case is thought to be advantageous.

In Chapter 5, I will consider the evolutionary significance of the mechanisms just listed. Right now, my point is simply to describe what they involve, not to comment on their frequency or importance. The point is to see how advantageousness and fitness may part ways. The fact that a trait would be good to have (or better to have than the alternative) does not mean that it has the higher fitness.

This decoupling of the concepts of fitness and advantageousness is an immediate consequence of how fitness is defined. Biologists don't really care about the fitness of single organisms—no one would bother to write a model about the fitness of Charlie the Tuna. What biologists care about is the fitnesses of *traits. The fitness of a trait is simply the average fitness of the organisms possessing it.* A given trait may be found in many organisms that differ among themselves in numerous ways. The fast zebras in our example may be resistant to disease or vulnerable to it; they may be good at digesting local grasses or not, and so on. The fast organisms have different fitnesses; the fitness of the trait is the average of these different values.

A consequence of this definition of trait fitness is that two traits found in precisely the same organisms must have the same fitness. With perfect correlation, the fast organisms *are* the organisms that are vulnerable to disease. If so, *Fast* and *Vulnerable* must have the same fitness value. But in spite of this commonality, we still can describe a difference between the two traits. Organisms survive *because* they are fast and *in spite of* the fact that they are vulnerable to disease. That is, there is *selection for* being fast but no *selection for* being vulnerable to disease.

To say that there is selection for one trait (*Fast*) and against another (*Slow*) is to make a claim about how those traits *causally contribute* to the organism's survival and reproductive success. On the other hand, to say just that one trait is fitter than another is to say nothing about *why* organisms with the first trait tend to do better than organisms with the second. One trait may be fitter than another because it confers an advantage *or* because it is correlated with other traits that do so.

Box 3.4 Hitchhiking and Intelligence

Darwin and the codiscoverer of the theory of evolution by natural selection, Alfred Russel Wallace, disagreed about the evolutionary origins of human intelligence (Gould 1980b). Wallace contended that natural selection cannot explain mental abilities that provide no practical benefits in surviving and reproducing. A keen eye is advantageous for hunting and gathering, but why should natural selection favor musical ability or the talent to invent novel scientific ideas? Wallace thought that natural selection can account for practical skills, not for higher capacities.

Darwin argued that natural selection can explain these higher capacities, even though these higher capacities were not useful to our ancestors. Rather, he thought that higher capacities *hitchhiked* on lower ones. The abilities that helped our ancestors solve practical problems crucial to survival were correlated with abilities that now help us solve theoretical problems that have no practical consequences at all.

How is the distinction between *selection of* and *selection for* applicable to this dispute?

When *Fast* and *Vulnerable* are perfectly correlated, the selection process will lead that combination of traits to increase in frequency. In the process, fast individuals get selected. Since the fast individuals are the ones that are vulnerable to disease, it also is true that the vulnerable individuals are selected. So two statements are true: there is *selection of* fast individuals, and there is *selection of* vulnerable ones. However, when we consider *why* the traits increased in frequency, the two traits cannot be cited interchangeably. There was *selection for* being fast, but there was no *selection for* being vulnerable to disease. "Selection for" describes the causes, while "selection of" describes the effects (Sober 1984b).

3.7 Teleology Naturalized

Biologists talk about the "functions" of various devices. For example, they say that the function of the heart is to pump blood. What could this mean? After all, the heart does many things. It pumps blood, but it also makes noise and takes up space in our chests. Why say that its function is to pump blood, rather than to make noise or to take up space? To understand claims about functions, we must clarify which of the effects that a device has is part of its function.

Perhaps the concept of function is clearest when we apply it to artifacts. We have no trouble discerning the function of a knife because knives are created and used with certain intentions. People make knives so that other people will be able to use them to cut. Of course, people have further motives (e.g., the profit motive) when they manufacture knives, and people can use knives for other purposes (e.g., as status symbols). These complications allow a knife manufacturer to say, "The function of this knife is to corner the market." And a king may say, "The function of this knife is to represent my authority." But notice that these remarks have something in common: Whether we say that the function of a knife is to cut or to make a profit or to represent authority, the claim is true because of the intentions that human agents have.

This raises the question of what it could mean to apply the concept of function to objects that are not the products of human handiwork. If organisms were the result of intelligent design, then the heart could be understood in the same format as the knife. To talk about the function of the heart would be to talk about the intentions that God had when he gave us hearts. But if we wish to give a purely naturalistic account of the living world, how can the idea of function make any literal sense? Perhaps it involves an unacceptable anthropomorphism, a vestige of a bygone age in which living things were thought of as products of intelligent design.

This suspicion—that functional concepts should be purged from biology—is encouraged by the fact that the scientific revolution in the seventeenth century eliminated teleology from physics. Aristotle's physics, like the rest of his view of nature, was saturated with teleology. He believed that stars, no less than organisms, were to be understood as goal-directed systems. An inner *telos* drives heavy objects to fall toward the place where the earth's center is. Heavy things have this as their function. Newtonian physics made it possible to think that a meteor may simply not have a function; it behaves as it does because of its conformity to scientific law. Talk of functions and goals is quite gratuitous. Perhaps progress in biology requires a similar emancipation from functional notions.

Darwin is rightly regarded as an innovator who advanced the cause of scientific materialism. But his effect on teleological ideas was quite different from Newton's. Rather than purge them from biology, Darwin was able to show how they could be rendered intelligible within a naturalistic framework. The theory of evolution allows us to answer the two conceptual questions about function posed before. It makes sense of the idea that only some of the effects of a device are functions of the device ("the function of the heart is to pump blood, not to make noise"). The theory also shows how assigning a function to an object requires no illicit anthropomorphism; it does not require the pretense that organisms are artifacts.

There is some variation in how evolutionary biologists use terms like "function" and "adaptation," but certain key distinctions are widely recognized. Seeing these distinctions is crucial; how we label them is less important.

To say that a trait is an "adaptation" is to comment not on its current utility but on its history. To say that the mammalian heart is (now) an adaptation for pumping blood is to say that mammals now have hearts because ancestrally, having a heart conferred a fitness advantage; the trait evolved because there was selection for having a heart, and hearts were selected because they pump blood. The heart makes noise, but the device is not an adaptation for making noise: The heart did not evolve *because* it makes noise. Rather, this property evolved as spin-off; there was *selection of* noise makers but no *selection for* making noise. More generally, we can define the concept of *adaptation* as follows:

> Characteristic *c* is an adaptation for doing task *t* in a population if and only if members of the population now have *c* because, ancestrally, there was selection for having *c* and *c* conferred a fitness advantage because it performed task *t*.

A trait may now be useful because it performs task *t*, even though this was not why it evolved. For example, sea turtles use their forelegs to dig holes in the sand, into which they deposit their eggs (Lewontin 1978). The legs are useful in this regard, but they are not adaptations for digging nests. The reason is that sea turtles possessed legs long before any turtles came out of the sea to build nests on a beach.

Conversely, an adaptation can lack current utility. Suppose wings evolve in some lineage because they facilitate flight. This means that wings are adaptations for flying. The environment then may change so that flying is actually deleterious—for example, if a new predator comes along that specializes on aerial prey. In this case, the wing is still an adaptation for flying, even though flying now diminishes an organism's fitness.

It follows that *adaptation* and *adaptive* are not interchangeable concepts. A trait is adaptive now if it currently confers some advantage. A trait is an adaptation now if it currently exists because a certain selection process took place in the past. The two concepts describe different temporal stages in the trait's career—how it got here and what it means for organisms who now have it. A trait can be an adaptation now without currently being adaptive. And it can be adaptive now, although it is not now an adaptation (for example, if it arose yesterday by mutation).

The concept of adaptation is sometimes used in a slightly more inclusive way. A trait is called an adaptation for performing some task if it either became common or remained common because it performed that task. Here, adaptation is applied to cover both the initial evolution and the subsequent maintenance of the trait.

In evolution, traits that evolved for one reason frequently get co-opted to perform some quite different task. For example, the penile urethra originally evolved because it was a conduit for urine; only subsequently did it become a conduit for sperm. Perhaps the trait is now maintained because it is a conduit for both. If we use the concept of adaptation in the extended sense just described, then the structure is an adaptation for both tasks. If we use the concept in its narrower sense, the penile urethra is an adaptation for one task but not the other.

I won't take a stand on which definition is "really" correct. Both are clear enough; the one we adopt is a matter of convenience. I will opt for the narrower definition. The important point is that on either the narrower or the broader definitions just cited, adaptation is a historical concept. Whether we are describing why the trait first became common or why it subsequently was maintained in the population, we are speaking in the past tense. Adaptation is not the same as current utility.

In addition to distinguishing the idea that a trait is an *adaptation* from the idea that the trait is adaptive, we also need to draw a distinction within the concept of adaptation itself. "Adaptation" can name a *process;* it also can name a *product*. The evolution of the wing involves the process of adaptation; the resulting wing is the product of that process. With respect to the process of adaptation, we need to distinguish *ontogenetic adaptation* from *phylogenetic adaptation*. An organism capable of learning is able to adapt to its environment. It modifies its behavior. A rabbit, for example, may learn where the foxes live and thereby avoid going to those places. Here, a change takes place during the organism's lifetime. The organism changes its behavior and thereby benefits.

The process of adaptation discussed in evolutionary theory is phylogenetic, not ontogenetic. Thus, protective coloration may evolve in a rabbit population because camouflaged rabbits avoid predators more successfully than uncamouflaged ones do. In this process of adaptation, no individual rabbit changes color—rabbits are not chameleons. Yet, natural selection modifies the composition of the population.

In the process of ontogenetic adaptation, it is easy to say who (or what) is adapting. The individual rabbit changes its behavior, and the rabbit obtains a benefit by doing so. But in the process of phylogenetic adaptation, who (or what) is doing the adapting? When protective coloration evolves, no individual rabbit is adapting since no individual organism is changing. Should we say that the population is adapting because the population's composition is changing? This often will be misleading since populations often evolve for reasons that have nothing to do with whether the changes will benefit them. As will be discussed in Chapter 4, natural selection usually is thought to favor traits because they benefit organisms, not because they happen to benefit groups. Protective coloration evolved (we may suppose) because it was good for individual rabbits, not because it was good for the group.

So far, I have said nothing about how the concept of *function* should be understood. Philosophers writing on this topic divide into two camps. There are those, like Wright (1976), who treat biological function in the way I have characterized adaptation; for them, to ascribe a function to some device is to make a claim about why it is present. For traits of organisms, assignments of function make reference to evolution by natural selection. And when we ascribe a function to an artifact, we are describing why that artifact was invented or kept in circulation. This is the *etiological view of functions*. Assignments of function are said to be hypotheses about origins.

The other philosophical camp rejects the idea that function should be equated with adaptation. For example, Cummins (1975) maintains that to ascribe a function to some device is not to make a claim about why the device is present. A function of the sea turtle's forelimbs is to dig nests, even if this is not why turtles have forelimbs. For Cummins, the limbs have this function because forelimbs contribute to some larger capacity of the organism.

One criticism raised against the etiological view is that biologists of the past competently assigned functions to various organs without ever having heard of the theory of evolution. William Harvey realized in the seventeenth century that the function of the heart is to pump blood. Antietiologists maintain that Harvey was making a claim about *what the heart does*, not about *why we have hearts*.

Another criticism of the etiological view is that it generates some odd consequences. Boorse (1976) describes a man who fails to exercise because he is obese. His obesity persists because he fails to exercise. Yet, it seems odd to say that the function of his obesity is to prevent exercise. This suggests that it is a mistake to equate function claims with explanations for why a trait is present.

On the other side, Cummins's theory has been criticized for being too permissive in the function ascriptions it endorses. The heart has a given weight. It contributes to the overall capacity of the organism to tip the scales at some number of pounds. Yet, it seems strange to say that a function of the heart is to weigh what it does. The trouble is that the distinction between function and mere effect seems to get lost in Cummins's theory. Every effect that an organ has can be counted as one of its functions, if we are prepared to consider the way that effect impacts on the organism as a whole.

There are other theories of function beyond the two just sketched, but I will not attempt to adjudicate among them here. Perhaps we should view the concept of adaptation as defined here as the one firm rock in this shifting semantic sea. If function is understood to mean adaptation, then it is clear enough what the concept means. If a scientist or philosopher uses the concept of function in some other way, we should demand that the concept be clarified. *Function* is a concept that should not be taken at face value.

The term "function" is often on the lips of biologists. However, this does not mean that it is a theoretical term in some scientific theory. "Function" is not like "selection coefficient" or "random genetic drift." It is used to talk *about* theories, but it does not occur ineliminably *in* any theory. Harvey discovered something important that the heart *does*. Unbeknownst to Harvey, the heart's behavior is a product of evolution. As long as we can speak clearly about current activities and their relationship to history, our descriptive framework will be on firm ground.

An interesting feature of all extant philosophical accounts of what the concept of function means is that they are *naturalistic*. Although the theories vary, they all maintain that functional claims are perfectly compatible with current biological theory. None requires that goal-directed systems possess some immaterial ingredient that orients them toward their appropriate end states. Whatever association teleology may have had with vitalism (Section 1.6) in the past, there is no reason why functional concepts cannot characterize systems that are made of matter and nothing else. The reason the concept of adaptation applies to organisms but not to meteors is not that living things contain immaterial ingredients. The difference derives from their very different histories. Selection processes cause some features of objects to be present because they conferred survival and reproductive advantages in the past. Other features are present for quite different reasons. This distinction can give meaning to the idea that function ascriptions apply to some characteristics of an object but not to others.

Suggestions for Further Reading

Rosenberg (1978, 1985) discusses the supervenience of fitness and other biological concepts, as does Sober (1984a). Mills and Beatty (1979) defend the propensity interpretation of fitness. Ayala (1974) draws useful distinctions among *ontological, methodological,* and *epistemological* reductionism. Brandon (1978) argues that the Principle of Natural Selection is the central principle of evolutionary biology and that it is untestable, although its instances are testable. Williams (1973) contends that the tautology problem can be solved by axiomatizing evolutionary theory and treating fitness as an undefined concept. Lewontin (1978) and Burian (1983) provide useful discussions of the concept of adaptation, and Sober (1984b) describes a selection toy that illustrates the difference between *selection of* and *selection for*. Mayr (1974) distinguishes *teleology* from *teleonomy* and explains how the latter concept is used in evolutionary biology.

4

The Units of Selection Problem

4.1 Hierarchy

We are organisms. For this reason, it may strike us as unproblematic and even obvious that genes and organs have the function of helping the organisms in which they are found. We find it natural to say that the heart has the function of keeping the organism alive; we describe genes as having the function of helping to construct useful bits of phenotype for the organism. When hearts or genes threaten the life of the organism, we say that they have *mal*functioned. It seems like a major departure from reality—the stuff of which science fiction is made—to invert this picture. Why not think of organisms as existing for the sake of the organs or genes they contain? Do hearts exist to serve us, or do we exist to serve our hearts? Do genes fulfill organismic functions, or are organisms simply survival machines that genes construct for their own benefit (Dawkins 1976)?

This was the problem that Samuel Butler (the nineteenth-century author of the Utopian novel *Erewhon*) gave voice to when he said that "a chicken is just an egg's way of making another egg." Do gametes have the function of producing organisms, or do organisms have the function of producing gametes? Or is it arbitrary to impose a functional asymmetry on the symmetrical fact that gametes produce organisms and organisms produce gametes?

The puzzle can be given a more general formulation. Let us take into account the fact that nature exhibits a multileveled hierarchy of organization. Consider the nested sequence of genes, chromosomes, cells, organs, organisms, local populations of conspecifics (*demes*), and multispecies communities of interacting demes. If organs have the function of helping organisms, is it also true that organisms have the function of helping the groups to which they belong?

This problem about *function* becomes more precise when it is formulated in terms of the concept of *adaptation*. Choose the objects at any level in the hierarchy; for example, consider organisms. To say that a property of an organism is an adaptation is to say that it evolved because there was selection for that property (Section 3.7). Now let us attend more closely to *why* there was selection for the trait. Did the property evolve because it benefited the *organisms* that had the trait, because it benefited the *groups* in which the trait was exemplified, or because it benefited objects at some other level still? The same question can be posed about objects at other levels. Groups have various properties that evolved because of natural selection. Did these evolve because they helped *groups* avoid extinction, as a consequence of what is good for the *organisms* in the group, or for some other reason?

We have here the basic issue that has come to be called the problem of *the units of selection*. I'll define this concept for two special cases and then in general:

> The organism is a unit of selection in the evolution of trait *T* in lineage *L* if and only if *T* evolved in *L* because *T* conferred a benefit on organisms.
>
> The group is a unit of selection in the evolution of trait *T* in lineage *L* if and only if *T* evolved in *L* because *T* conferred a benefit on groups.
>
> *X* is a unit of selection in the evolution of trait *T* in lineage *L* if and only if *T* evolved in *L* because *T* conferred a benefit on Xs.

A trait that evolved because it benefited the organisms that possessed it is an organismic adaptation; if it evolved because it benefited the groups in which it was found, then it is a group adaptation. The units of selection problem is the problem of determining what kinds of adaptations are found in nature.

Two logical features of these definitions are worth noting. The first is that the units of selection issue concerns evolutionary history, not current utility. Groups may now possess various traits that help them avoid extinction, but it is a separate issue whether those traits evolved *because* they have that effect. If they evolved for some other reason, then those traits provide a *fortuitous group benefit;* they are not *group adaptations*. (This point connects with the distinction drawn in Section 3.7 between saying that a trait is an *adaptation* and saying that it is *adaptive*.)

The second logical point is that the definition allows that different traits may have evolved for different reasons and that a single trait may have evolved for several reasons. Perhaps some traits are organismic adaptations while others are group adaptations. In addition, it is possible for a given trait to evolve because it simultaneously benefits objects at several levels of organization. This means that the question "What is the single unit of selection in all of evolution?" may have a false presupposition. Even if there is a single unit of selection in all the selection processes that have ever occurred, this should not be assumed in our formulation of the problem but should be argued for explicitly.

The question raised by Butler's puzzle about the chicken and the egg can be transposed to the contrast between group and individual adaptation: Is there an objective difference between saying that a trait is a group adaptation and saying that it is an organismic adaptation? Why isn't it a matter of convention whether one describes a trait as evolving for the good of the organism or for the good of the species?

The answer to this question can be given in one word: *altruism*. An altruistic trait is one that is deleterious to the individual possessing it but advantageous for the group in which it occurs. If the organism is the exclusive unit of selection, then natural selection works *against* the evolution of altruism. If the group is sometimes a unit of selection, then natural selection sometimes *favors* altruistic traits. The units of selection problem cannot be settled by stipulative convention, because different views about the units of selection make contrary predictions about which traits evolve under natural selection. The important point is that there can be *conflicts of interest* between objects at different levels of organization: What is good for the group may not be good for the organism.

To clarify this idea, consider two examples of traits that seem, at first glance, to be altruistic. Honeybees disembowel themselves when they sting intruders to the nest. They thereby sacrifice their own lives and help the group to which they belong. Similarly, crows often issue warning cries when a predator approaches. This behavior seems to put sentinels at a disadvantage; the other members of the group receive a benefit while sentinels call the predator's attention to themselves. Of course, the costs and benefits may not be as they seem. If the warning cry is hard for the predator to localize or if the alarm call actually protects the sentinel by sending the rest of the flock into a frenzy of activity, then the sentinel behavior may not actually be altruistic. But waiving these complications for the moment, we can see the barbed stinger of a honeybee and the sentinel cry of a crow as traits that seem to be bad for the organisms possessing them but good for the group in which they occur.

In offering these traits as *prima facie* examples of altruism, I am not saying that bees or crows have altruistic psychological motives. The evolutionary concept of altruism concerns just the fitness effects, to self and other, of the behavior involved. Thus, plants and viruses can be altruistic even though they do not have minds.

If organisms compete against other organisms within the confines of a single population, then natural selection will favor selfish organisms over altruistic ones. Selfish individuals are "free riders"; they receive the benefits of altruistic donation without incurring the cost of making donations themselves. On the other hand, if groups compete against other groups and if groups of altruists do better than groups of selfish individuals, then altruism may evolve and be maintained. Organismic adaptations evolve by a process of organismic selection; group adaptations evolve by a process of group selection. The units of selection problem, since it concerns the kinds of adaptations found in nature, has to do with the kinds of selection processes that produced the traits we observe.

The units of selection problem has been with us ever since Darwin. Although he is sometimes said to have held that traits evolve because they are "good for the species," Darwin almost never thought about natural selection in this way. For example, consider the position that he took in his disagreement with Wallace about the proper explanation of hybrid sterility (Ruse

1980). If hybrids (the result of matings between individuals in different species) were usually sterile or inviable, it would make sense for organisms to restrict their mating to conspecifics. But given that matings across species boundaries do occur, how is one to explain the fact that the progeny of such matings are often sterile? Clearly, it is not in the interest of the hybrid organism to be sterile, nor do the parents gain an advantage by having offspring that cannot reproduce. So how is hybrid sterility to be explained?

Wallace argued that the trait is a species-level adaptation; he contended that it is advantageous to the species that hybrids be sterile. That way, the species does not blend with other species; it thereby retains its own distinctive characteristics. For Wallace, hybrid sterility evolved because it was good for the species and in spite of the fact that the trait was not good for the sterile hybrids themselves.

Darwin rejected this argument; he preferred the hypothesis that hybrid sterility is not an adaptation for anything. It is *spin-off* (Section 3.6)—a by-product of selection acting within each of the species considered. Species *X* evolves one suite of phenotypes and species *Y* evolves another. In each case, the morphological, physiological, and behavioral traits evolve because of the advantages they confer on organisms. A consequence of the separate evolutionary processes occurring within each species is that the organisms in the two species become quite different. This means that between-species matings rarely occur, that they rarely produce viable offspring, and that the offspring produced are rarely fertile. Perhaps hybrid sterility provides a fortuitous benefit to the species; it is not a species-level adaptation.

Almost all of Darwin's many selectionist explanations deploy the concept of individual, not group, selection. However, there are a few contexts in which Darwin forsakes an individualistic interpretation of adaptation. One of them occurs in the *Descent of Man* in his discussion of human morality. Here is Darwin's statement of the problem (p. 163):

> It is extremely doubtful whether the offspring of the more sympathetic and benevolent parents, or of those which were the most faithful to their comrades, would be reared in greater numbers than the children of selfish and treacherous parents of the same tribe. He who was ready to sacrifice his life, as many a savage has been, rather than betray his comrades, would often leave no offspring to inherit his noble nature. The bravest men, who were always willing to come to the front in war, and who freely risked their lives for others would on average perish in larger numbers than other men.

If altruistic self-sacrifice is deleterious for the individual, though good for the group, how can it evolve? Here is Darwin's answer (p. 166):

> It must not be forgotten that although a high standard of morality gives but a slight or no advantage to each individual man and his children over the other men of the same tribe, yet that an advancement in the standard of morality and an increase in the number of well-endowed men will certainly give an immense advantage to one tribe over another.

Within any particular group, altruists do worse than selfish individuals, but groups of altruists do better than groups of selfish individuals. Altruism can evolve by group selection.

Another place where Darwin seems to stray from the straight and narrow of individual selection is his discussion of sterile castes in the social insects. This was a problem that mattered a great deal to Darwin as he worked toward his theory of natural selection (Richards 1987). Here is what he says about it in the *Origin* (p. 236, emphasis added):

> How the workers have been rendered sterile is a difficulty, but not much greater than that of any striking modification of structure; for it can be shown that some insects and other articulate animals in a state of nature occasionally become sterile; and if such insects had been social, and it had been profitable *to the community* that a number should have been annually born capable of work, but incapable of procreation, I can see no very great difficulty in this being effected by natural selection.

Sterility is disadvantageous to the organism, but groups that contain sterile workers may do better than groups that do not.

Having sketched some of Darwin's discussion of the problem of the units of selection, I'll now jump ahead fifty years or so to the heyday of the Modern Synthesis. The population geneticists who helped forge the Modern Synthesis were fairly skeptical about the evolution of altruistic characters. Fisher (1930) thought that, with the possible exception of the evolution of sex, the good of the group is the wrong way to think about why adaptations evolve. Haldane (1932) also was dubious about altruism, and so was Wright (1945). None of these authors suggested that it is *impossible* for an altruistic characteristic to evolve and be maintained, but their basic outlook was that the circumstances needed for this to happen are exemplified in nature only rarely. Most of what we see in nature, they said, can be understood without the concepts of group selection and group adaptation.

The arguments advanced by these population geneticists were based on thinking about simple quantitative models. However, during the same period, there were several influential ecologists and field naturalists who came at the problem from a rather different angle. Their detailed empirical examination of natural populations led them to think that they had *observed* organisms with altruistic characteristics. These observations, they thought, demand explanation in terms of the idea of group selection. Allee, Emerson, Park, Park, and Schmidt (1949, p. 728) built a detailed case for the view that

> natural selection operates on the whole interspecies system, resulting in the slow evolution of adaptive integration and balance. Division of labor, integration, and homeostasis characterize the organism and the supraorganismic interspecies population. The interspecies system has also evolved these characteristics of the organism and may thus be called an ecological superorganism.

Wynne-Edwards (1962) developed a similar line of argument. He thought that many long-standing populations avoid overexploiting food resources and also avoid overpopulation. If the organisms in a population have too many offspring and drastically deplete supplies of food, the population will crash to extinction. An organism that restrains itself is altruistic; it reduces its own fitness and thereby makes the population better off.

This same pattern of thinking is found in much of the work in ethology during this period. For example, Lorenz's (1966) explanation for why conspecifics do not battle to the death was that this is an adaptation that exists to preserve the species. He assumed that a purely selfish individual would show no such restraint. Symbolic combat, therefore, is a kind of altruism; it is good for the group though deleterious to the individual.

It is interesting that quantitative model builders and field naturalists so often reached opposite answers to the question of altruism. To this day, it remains true that many biologists see the world either through the lens of mathematical models or through a detailed knowledge of the biology of some group of organisms. If mathematical models predict that altruism rarely exists, two reactions are possible. One response is to conclude that the mathematical models are misguided. The other is to claim that what seems to be altruism really is not because the models must be taken seriously.

During the period 1930–1962, the two sides in this dispute rarely made contact with each other. Population geneticists had more pressing problems than the issue of altruism, and field naturalists often felt that there was no special need for them to develop quantitative models of the processes they postulated. This all changed in 1966 with the publication of George C. Williams's *Adaptation and Natural Selection*. Williams developed his argument in English prose, with nary an equation in sight. His main point was that group selection hypotheses were the result of sloppy thinking. It wasn't that group adaptation is implausible because of some recondite observation that had only recently come to light. The problem wasn't so much empirical as conceptual. Simple and fundamental facts about the way natural selection should be understood render group adaptation a concept *non grata*.

A few years before Williams's book appeared, William Hamilton published a pair of groundbreaking papers on the evolution of social behavior. He showed how cooperative behavior could be to the advantage of the cooperators. Hamilton (1964) argued that although donation involves a sacrifice of the donor's Darwinian fitness, there is another concept—"inclusive fitness"—that organisms can augment by judicious donation. Hamilton's papers were viewed as undercutting the idea that apparently altruistic behaviors are genuinely altruistic.

In Williams's (1966) book, he notes that a female-biased sex ratio would be *prima facie* evidence of group selection and adaptation. Recall from Section 1.4 that Fisher's argument predicted equal investment in the two sexes: If sons and daughters involve equal costs, this entails that an equal number of males and females should be produced. Williams takes Fisher's argument to heart; although it would be good for the group to have more females than males, Williams thought that a purely individual-level account would predict an even sex ratio.

It turns out that many arthropods have female-biased sex ratios. But Hamilton's (1967) paper a year later argued that "extraordinary sex ratios" could be understood in terms of the individual advantages they provide. Once again, what could have been an observation favoring the hypothesis of group adaptation was reinterpreted as an observation that requires no such explanation.

In the years following Williams's (1966) book, a number of biologists investigated various mathematical models of group selection. They asked how easy or hard it would be for an altruistic character to evolve and be maintained. The conclusion they reached, critically reviewed in Wade (1978), was that altruism can evolve only in a relatively narrow range of parameter values. It isn't that altruism is *impossible;* rather, they concluded, altruism is relatively *improbable.*

Since then, the idea of group selection has refused to die completely, although it is still a concept that many biologists are reluctant to consider. David Wilson's (1980) book attempted to model group selection in a way that has real empirical application. And those working in the tradition of Sewall Wright's (1931) models of interdemic selection have kept the idea alive. In addition, a number of paleobiologists have developed the idea of *species selection,* according to which patterns of diversity among species cannot be explained solely on the basis of individual selection but require a nonrandom sorting process at a higher level (Eldredge and Gould 1972; Stanley 1979; Vrba 1980; discussed in Sober 1984b, Section 9.4).

In the remainder of this chapter, I will not try to assess the empirical evidence for and against the idea of group adaptation. Rather, I will try to make clear what kind of empirical question the units of selection problem poses. This is worth doing because it is all too easy to define altruism as "what cannot evolve" and selfishness as "what must evolve." This makes the problem of group adaptation look like a nonissue. For those attracted by this definitional sleight of hand, it should be puzzling why so many very smart biologists, from Darwin's time down to the present day, have thought that the units of selection problem has biological substance. If the idea of group adaptation is to be rejected, it should be rejected for the right reasons.

Williams's (1966) brilliant critique of group adaptation was not entirely negative. Besides saying what he rejected, he also presented a position that he thought was correct. Although Williams saw himself as defending Darwinism, he did not advocate a return to Darwin's usual idea that the organism is the unit of selection. Instead, he argued that it is not the group nor even the organism that is the unit of selection. For Williams, the unit of selection is the "meiotically dissociated gene." This is the idea that Dawkins (1976) subsequently popularized in his book *The Selfish Gene.* In light of Williams' suggestion, we will have to expand our list of the alternative positions that might be taken on the units of selection problem. The choices are not limited to *the group* versus *the organism;* they need to include *the gene* as well.

As stated before, the welfare of the group can conflict with the welfare of the organism; when this is true, we use the concept of altruism to describe the situation. I will claim in what follows that the interests of the organism can conflict with the interests of the gene. Just as groups are made of organisms, so we can think of organisms as made of genes. Conflicts of interest are possible between objects at different levels.

Here, I am treading on controversial territory. Everyone grants that what may be good for the group may not be good for the organism. However, Dawkins (1976) thinks of the genic and the organismic points of view as equivalent. I will argue otherwise. But there is more. Some of the arguments that Williams and Dawkins present for the thesis of *genic selectionism*—that the gene is the unit of selection—have a curious characteristic. They cite facts about the evolutionary process that are quite compatible with group selection and group adaptation. If genic selectionism really is incompatible with group selection, then an argument for the former should cite facts that count against the latter. As we will see, something has gone seriously wrong in these arguments.

Before we get to the selfish gene, it is well to begin our investigation where the units of selection problem had its historical origin. We need to see more clearly how group selection and adaptation differ from individual selection and adaptation.

4.2 Adaptation and Fortuitous Benefit

In Section 3.1, I stated a simple rule of thumb for saying when a trait will increase in frequency. If natural selection is the only force influencing a heritable trait's evolution, then *fitter traits increase in frequency and less fit traits decline*. Note that this criterion is stated in terms of *relative fitness*; it concerns which trait is fitt*er*. I now will explain why this rule of thumb leaves entirely open the question of whether the average fitness of the organisms in the population goes up as the population evolves. *Absolute fitness* need not improve under selection.

Consider the simple selection story discussed before. Natural selection favors fast zebras over slow ones because fast zebras are better able to avoid predators. We will adopt the simplest of assumptions about heredity—suppose that zebras reproduce uniparentally and that offspring always resemble their parents. These assumptions entail that if a population is composed of slow zebras and a fast mutant or migrant is introduced, the novel trait will increase in frequency. If we ignore the complication that other forces (like drift) may influence the trait's evolution, we can predict that the fast trait will go all the way to fixation (100 percent).

A graphical representation of this selection process is given in Figure 4.1. This representation says that the fitnesses of the two traits are unaffected by their frequencies. Fast individuals have a given probability of surviving to adulthood (or a given expected number of offspring), and slow individuals have another, lower, fitness; the two fitnesses are *frequency independent*.

In this selection process, while *Fast* is supplanting *Slow*, it also is true that the average fitness of the organisms in the population ($\overline{w}$) increases. By the end of the process, all the zebras are fitter than the zebras were before the process began.

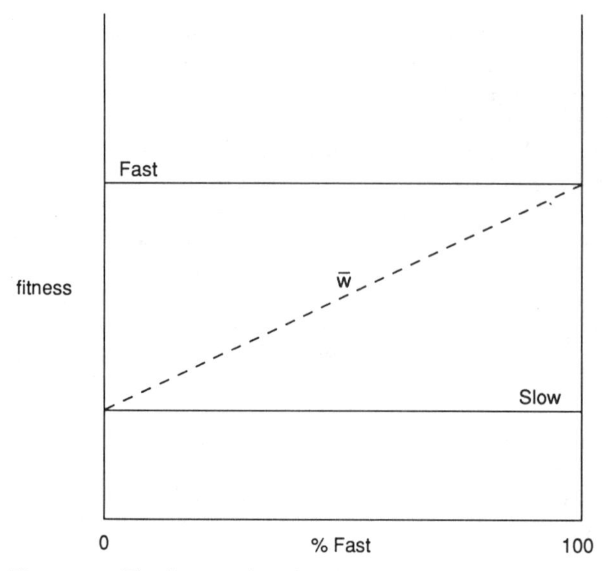

Figure 4.1 The fitness values for *Fast* and *Slow* are independent of the traits' frequencies in the population. As *Fast* increases in frequency, the average fitness of the organisms in the population ($\overline{w}$) also increases.

If $\overline{w}$ measures the average fitness of a zebra *in* a group, it also is natural to regard $\overline{w}$ as measuring the fitness *of* the group. A group of slow zebras has a certain chance of going extinct (e.g., by having all its member organisms devoured by lions). A group of fast zebras has a lower chance of being destroyed. So the process increases both the fitness of the individuals in the group and the fitness of the group itself.

Although these are both effects of the selection process, suppose we now ask *why* the fast trait went to fixation. Did it evolve because it was advantageous to the organisms possessing it or because it was good for the group as a whole? The answer is that increased running speed evolved because it benefited the organisms. Speed is an individual adaptation; the fact that the group is better off just shows that increased running speed provides a fortuitous group benefit. Running fast is not a group adaptation.

This point has a great deal of generality to it: When the organisms in a population evolve under the influence of individual selection, what determines the evolution of the population is the relative fitness of organisms; the effect on the fitness of the group has nothing to do with how the system evolves.

To see this, consider an example described by Lewontin (1978). A population is at its carrying capacity; its census size is as large as its suite of characters and its environment allow. Suppose a mutant is introduced that produces twice as many eggs as the resident phenotypes do. This trait will sweep to fixation. However, by the end of the process, the population will have the same

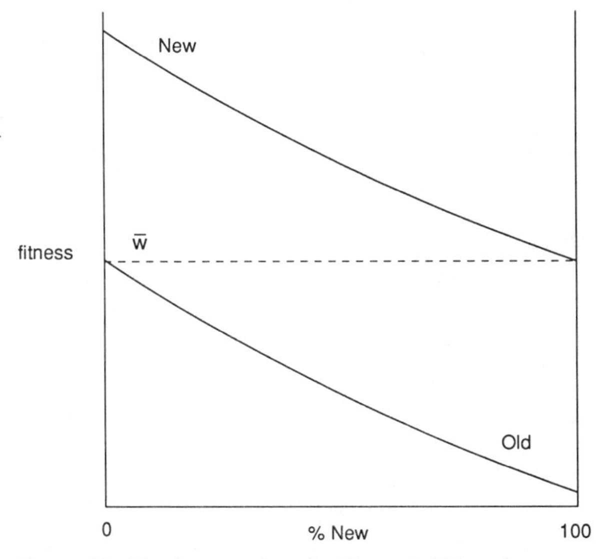

Figure 4.2 The fitness values for *New* and *Old* are frequency dependent. As *New* increases in frequency, the average fitness of the organisms in the population ($\overline{w}$) remains constant.

census size that it had at the beginning. Before the novel trait appeared, individuals were reproducing at replacement levels; after the novel trait sweeps to fixation, the same level of productivity prevails. This process is illustrated in Figure 4.2. Note that $\overline{w}$ has the same value at the end of the process that it had at the beginning.

As a final example, consider the two traits (S and A) depicted in Figure 4.3. What will happen if trait S is introduced into a population of A individuals? Since S is always fitter than A, S increases in frequency all the way to fixation. But notice that $\overline{w}$ goes downhill. The population is worse off at the end of the process than it was at the beginning.

Consider a hypothetical example. In a given population, there are organisms that pollute the environment and organisms that do not. Suppose that pollution is bad for everyone, but that it harms nonpolluters more than it harms polluters. In this circumstance, the polluting trait will increase in frequency and eventually will become universal. In the process, the organisms in the population do worse and worse; as the level of pollution increases, the population may even drive itself to extinction.

One often thinks of natural selection as an *improver*; fitter traits replace less fit traits, and the organisms at the end of the process are fitter than the organisms were at the beginning. Figure 4.3 shows that this need not be so. Selection *can* improve the average fitness of organisms. However, this is not inevitable.

The traits in Figure 4.3 were called S and A for a reason. This figure represents two simple facts about the relationship of evolutionary selfishness and

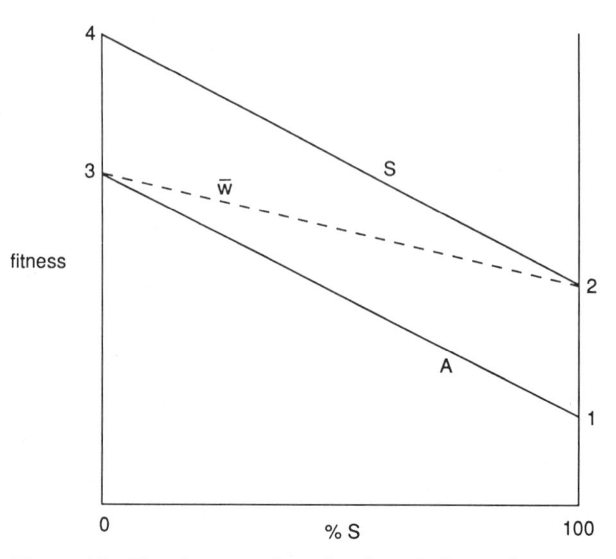

Figure 4.3 The fitness values for *S* and *A* are frequency dependent. As *S* increases in frequency, the average fitness of the organisms in the population ($\overline{w}$) declines.

altruism. Take an ensemble of populations, each with its own local mix of selfish and altruistic organisms. Within any population, selfish individuals are fitter than altruists. But altruistic populations have a higher fitness than selfish populations. These two ideas provide the basis for the two-level process needed for altruism to evolve. Within each population, individual selection favors selfishness over altruism. But there is competition among populations, and this favors altruism over selfishness. The final outcome depends on the strengths of these two conflicting forces.

The representation of selfishness and altruism in Figure 4.3 entails that altruism is not the same thing as helping behavior. For example, parental care is not altruistic if parents who help their offspring are fitter than parents in the same group who do not. Don't forget that an organism's fitness includes both survival *and* reproduction. Maybe sentinel crows and kamikaze bees are altruists, but caring parents are not. To see whether a trait is altruistic, one must compare its fitness with the fitnesses of the other traits with which it competes.

4.3 Decoupling Parts and Wholes

A defining characteristic of evolutionary altruism is that altruists are less fit than selfish individuals within the same group. The rule of thumb we use to determine a population's trajectory is that fitter traits increase in frequency and less fit traits decline. These two ideas seem to combine in syllogistic form to tell us that altruism cannot evolve:

Within any group, altruism is less fit than selfishness.
Less fit traits decline in frequency.

———————————

So, altruism will decline in frequency.

What is true enough is that altruism cannot evolve *if the selection process takes place within the confines of a single group*. But if the system under consideration is an ensemble of groups, the argument is fallacious. The definition of altruism and the rule of thumb do not entail that altruism must decline.

To see why, we must break the grip of a very powerful, though mistaken, commonsense idea about the relation of parts and wholes. We find it overwhelmingly natural to think that what is true in every part automatically must be true in the whole. If altruism is less fit than selfishness in each group, won't altruism automatically be less fit than selfishness in the whole ensemble of groups? The answer to this question is no.

To see why, I'll present a very simple example. I'll use Figure 4.3 to calculate the fitnesses of altruism and selfishness within each of two groups *and* within the entire two-group ensemble. Suppose the two populations contain 100 organisms each. Group 1 is 1 percent selfish. Group 2 is 99 percent selfish. The fitnesses and the number of organisms are as follows:

Group 1	*Group 2*	*Global average*
$1S$; $w = 4$	$99S$; $w = 2$	$100S$; $w = 2.02$
$99A$; $w = 3$	$1A$; $w = 1$	$100A$; $w = 2.98$

Notice that selfishness is fitter than altruism within each group ($4 > 3$ and $2 > 1$), but that this inequality reverses when we consider the global averages ($2.02 < 2.98$). What is true in each part is not true in the whole.

Things get stranger still if we track the evolution of this two-group ensemble over a single generation. I'll interpret the fitness values as expected numbers of offspring. Let us suppose that the organisms reproduce uniparentally, with like begetting like, and that the parents die after reproducing. Here is what happens to the frequencies of the two traits, both within each group and in the global ensemble:

	Group 1	*Group 2*	*Global ensemble*
Parental census	$1S$; $99A$	$99S$; $1A$	$100S$; $100A$
Parental frequencies	$1\%S$; $99\%A$	$99\%S$; $1\%A$	$50\%S$; $50\%A$
Offspring census	$4S$; $297A$	$198S$; $1A$	$202S$; $298A$
Offspring frequencies	$1.3\%S$; $98.7\%A$	$99.5\%S$; $0.5\%A$	$40\%S$; $60\%A$

Altruism declines in frequency within each group, but it increases in frequency in the global ensemble.

It is difficult to break the grip of the commonsense picture of how wholes and parts must be related. If Democrats are declining in frequency in every state of the Union and Republicans are increasing, doesn't it follow that Democrats are declining in the United States as a whole? The answer is that *this does not follow.*

The decoupling of wholes and parts that we see at work in this example manifests a more general idea, which statisticians call *Simpson's paradox*. Here is another example, this one quite unrelated to evolutionary problems. The University of California at Berkeley was suspected of discriminating against women in admission to graduate school (Cartwright 1979). Women were rejected more often than men, and the difference was big enough that it could not be dismissed as due to chance. However, when academic departments were examined one by one, it turned out that women were rejected no more frequently than men. To see how these two statistics are compatible, let us invent a simple example. Let 100 men and 100 women apply to two departments, each with its own acceptance rate (which is the same for applicants of both sexes):

	Department 1	*Department 2*	*Global*
Number applying	90 women; 10 men	10 women; 90 men	100 women; 100 men
Percent accepted	30%	60%	
Number accepted	27 women; 3 men	6 women; 54 men	33 women; 57 men

Although the acceptance rate for men and for women is the same in each department, women were admitted less often overall. What is true within each department is not true within the university as a whole.

The key to Simpson's paradox is *correlation*. In the first example, altruists tend to live with altruists. In the second example, women tend to apply to departments with lower acceptance rates. If the two groups in the first example contained the same mix of altruists, then altruism could not evolve; and if each academic department attracted the same mix of male and female applicants, women would be accepted to graduate school as often as men. A general characterization of Simpson's paradox is provided in the accompanying box.

The two-group example of the evolution of altruism should not be overinterpreted. It involves a one generation snapshot. I analyzed the system after it had reached 50 percent altruism, but I did not describe how altruism might have evolved that far. I then traced the process into the next generation but did not investigate the process in the longer term. So the example should not be taken to describe the whole process by which altruism can evolve by group selection. My point is more modest. This section began with a simple syllogism whose conclusion is that altruism cannot evolve. I hope that the example and the idea of Simpson's paradox show why this syllogism is fallacious. By definition, altruism is less fit than selfishness *within each group*; but if like tends to live with like, it may turn out that altruism has a higher global fitness than selfishness. If so, altruism will increase in frequency.

What will happen if we follow the example of the two-group ensemble for a number of generations into the future? If the two groups hold together, self-

Box 4.1 Simpson's Paradox

Simpson's paradox arises if two properties *C* and *E* are positively correlated within each of a set of subpopulations $B_1, B_2, \ldots, B_n$, but are not positively correlated when one averages over those subpopulations:

$P(E / C\&B_i) > P(E / -C\&B_i)$, for each *i*.
$P(E / C) \not> P(E / -C)$.

More generally, Simpson's paradox arises when >, =, or < occurs in the first condition but not in the second.

The accompanying text provides two examples of Simpson's paradox. In the second example, probabilities of admission to graduate school can be calculated within each academic department and over the entire university. What are the overall probabilities of admission to graduate school for women and for men?

The first example involves group selection. If the fitnesses in that example are interpreted as probabilities of surviving, the two conditions that define Simpson's paradox apply straightforwardly. However, the example was developed by understanding fitnesses as expected numbers of offspring. To see how this is an example of Simpson's paradox, replace *P(/)* with *Exp(/)* in the above formulation.

In Box 3.3, which defines the concept of correlation, an example about smoking, high cholesterol, and heart attacks is described. How does Simpson's paradox apply to that example?

ishness will continue to increase in frequency within each of them. In the limit, selfishness will go to fixation within each group, and so the global frequency of altruism must go to zero. Each group experiences what Dawkins (1976) called *subversion from within*.

For altruism to evolve, the groups must *not* hold together indefinitely. They must differ in their extinction and colonization rates. So let us imagine that groups fragment into small propagules that go forth to found colonies once the parent population has reached a given census size. Altruistic groups grow faster than selfish ones, so altruistic groups found colonies more often and go extinct less often than groups of selfish individuals.

Two further elements must be added to this picture if we are to see how altruism can evolve by group selection. We first must consider how a parent group influences the mix of altruistic and selfish individuals found in the offspring colonies that it founds. Suppose that a parent population sends out propagules when it reaches a census size of 1,000 organisms. Consider a population with that census size that happens to be 90 percent altruistic. If the parent population divides into 100 offspring propagules of 10 organisms each, what will be the mix of altruism and selfishness in these small propagules?

If altruism is to evolve, daughter colonies should *not* precisely resemble their parents. If all the daughter colonies begin life at 90 percent altruism, then by the time *they* reach a census size of 1,000, the percentage of altruism will have declined. The result will be a kind of ratcheting, with altruism declining in frequency in every generation. What is needed, instead, is a sampling process, which allows some daughter colonies coming from a 90 percent parent to begin life at higher than 90 percent altruism while others begin life with a lower frequency.

The other crucial requirement is *time*. Suppose that selfish individuals are sufficiently fitter than altruists that any population, no matter where it begins, will reach 100 percent selfishness within 20 generations. If this is true, then altruism cannot evolve if groups found colonies every 90 organismic generations. The colonization phase of the process will have come too late. For altruism to evolve, the group selection part of the process must occur at something like the rate at which individual organisms die and reproduce.

4.4 Red Herrings

I hope that the previous section provided an intuitive feel for the kinds of considerations that are relevant to the question of whether altruism will evolve. If evolution takes place within the confines of a single population, there is no variation among groups and so there can be no group selection. If there is variation among groups, but groups go extinct and found colonies very slowly, then the group selection part of the process may be too weak to counteract the spread of selfishness within each group. On the other hand, if groups vary in their mix of selfishness and altruism and if they die and reproduce at something approaching the time scale on which organisms die and reproduce, then altruism may be able to evolve. I'll summarize the relevant considerations with the following slogan: *Population structure is of the essence.*

This approach to the units of selection problem has implications for a number of arguments that have been given in the literature. We will see that these arguments bring up considerations that are entirely irrelevant to deciding whether traits evolve because they are good for the group.

The first such argument claims that the gene, rather than the group, is the unit of selection because the gene is the unit of heredity. Dawkins (1976, pp. 28, 33) puts forward this argument, which is also found in Williams (1966); Dawkins even went so far as to *define* what a gene is in such a way that the gene must be the unit of selection.

Dawkins (1976, p. 11) also says that the selfish gene idea is "foreshadowed" by August Weismann's idea of the continuity of the germ plasm. Weismann opposed the Lamarckian idea of the inheritance of acquired characteristics. For example, if a mother giraffe lengthens her neck by stretching, this does not allow her offspring to achieve long necks without needing to stretch. The Weismann doctrine is illustrated in Figure 4.4. Parents influence their offspring by way of the genes they transmit; those genes are not modified by the phenotypes that the parents acquired in their own lifetimes.

The dispute about group adaptation has nothing to do with the issue of whether Mendelism or Weismannism is true. Accepting the idea that the gene

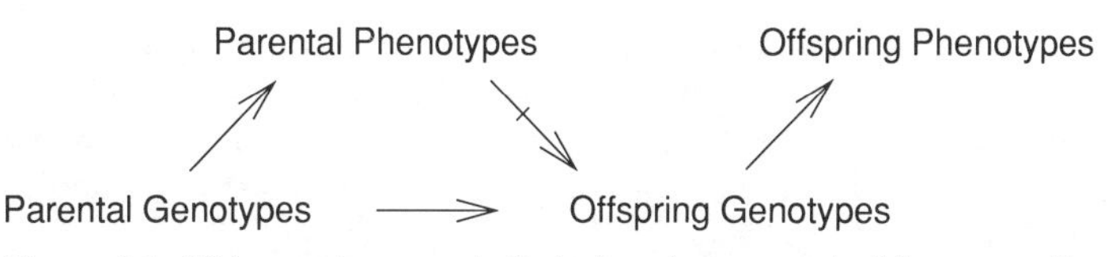

Figure 4.4 Weismannism asserts that phenotypes acquired in a parent's lifetime do not modify the genes that the offspring receives.

is the unit of heredity and that Lamarckian views about the relationship of genotype to phenotype are wrong says nothing about whether group adaptations are common, rare, or nonexistent.

Williams and Dawkins also argue that genic selectionism is correct on the grounds that all selection processes can be "represented" in terms of genes and their properties. The premiss of this argument can be interpreted so that it is true, but nothing whatever follows about whether there are group adaptations. If we define evolution as change in gene frequency, then it must be true that when evolution is caused by natural selection, some genes will be fitter than others. In this sense, it always is possible to talk about evolution by natural selection in terms of allelic frequencies and allelic fitnesses.

These points may be summarized by considering Figure 4.3. Earlier, I interpreted this figure as describing the relationship between two *phenotypes* (altruism and selfishness), but I could just as easily have said that *A* and *S* are two *alleles* that exist at a locus in a haploid organism. We could use the figure to describe the fitnesses and frequencies of the two genes, both within each group and across the ensemble of groups. Nothing prevents us from understanding these genes for altruism and selfishness as obeying the usual Mendelian and Weismannian rules. Such a genic description is certainly possible, but it does not rule out the possibility that group selection is causing group adaptations to evolve.

Williams (1966) and Dawkins (1976) also contend that the single gene is the unit of selection because genes have a *longevity* that gene complexes, organisms, and phenotypes do not. Socrates's phenotype occurred just once, but his individual genes were passed down the generations (in the form of copies). Again, the point to notice is that the longevity of genes does nothing to undercut the idea of group selection. Even if "genes are forever," this leaves open the question of whether altruistic genes evolve by group selection.

These arguments for genic selectionism have a common defect. They seize on facts about evolution that will be true *whether group selection occurs or not*. This means that proponents of genic selectionism face a dilemma. If all they mean by their position is that Mendelism is true or that Weismannism is true or that evolution can be described in terms of what happens to genes, then their position is trivial. It masquerades as an alternative to the idea of group selection, but it is really no such thing (since Mendelism and Weismannism and gene level descriptions are quite compatible with group selection). On the other hand, if the selfish gene idea really is a competitor with the idea of group selection, then these arguments should be set aside as *non sequiturs*.

Another spurious argument about the units of selection problem concerns the causal chain that leads from an organism's genotype to its phenotype to its survival and reproduction:

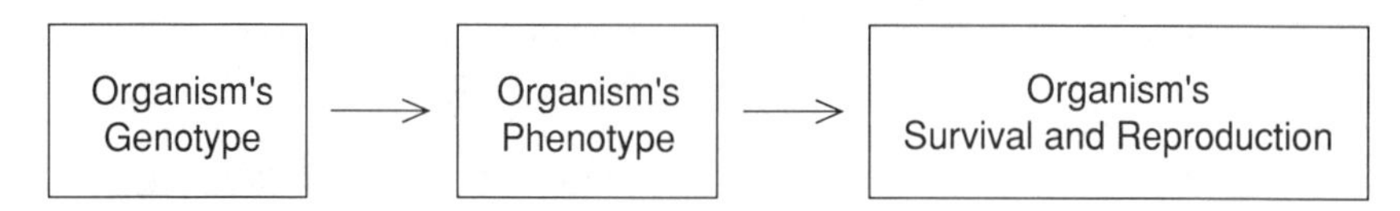

Mayr (1963, p. 184) and Gould (1980b, p. 90) have emphasized that natural selection acts "directly" on the organism's phenotype and only "indirectly" on its genes.

Gould contends that this asymmetry shows that the gene is not the unit of selection. But, in fact, Williams (1966) and Dawkins (1976) actually *embrace* the idea depicted in this causal chain. For Dawkins especially, genes are the deeper causes of what happens in evolution; they build survival machines (i.e., organisms), which are constructed to promote the interests of the genes they contain.

The causal chain just described does not discredit genic selectionism. But neither does it show that it is correct. Once again, group selectionists can accept the idea that the genes inside an organism cause the organism to be altruistic or selfish and that the organism's phenotype influences its fitness. Of course, these concessions should not be taken to deny the relevance of the *environment* at each stage. The organism's phenotype is influenced by both its genotype *and* its environment. And its fitness is influenced both by its phenotype *and* by the environment it inhabits. In any event, this causal chain says nothing about whether group adaptations (and the genes that code for them) have evolved by group selection.

Another argument that Williams (1966) and Dawkins (1976) present appeals to the idea that lower-level selection hypotheses are more *parsimonious* than hypotheses of group selection. Here, Williams is appealing to the methodological maxim called *Ockham's razor*: If two explanations can both explain the observations at hand, we should prefer the explanation that postulates fewer entities or processes or that makes the smallest number of independent assumptions. Parsimonious theories are "tightfisted" in what they say. Williams argues that this general methodological consideration provides a reason for rejecting a group selection hypothesis if an explanation in terms of individual selection can be constructed.

Williams uses this type of argument in his discussion of why musk oxen "wagon-train" when they are attacked by predators: When attacked by wolves, the males form a circle while the females and young shelter in the circle's interior. The behavior of the males is at least *prima facie* an example of altruism since males protect individuals other than their own progeny. A group selection hypothesis might seek to explain this behavior by saying that groups that wagon-trained fared better in the struggle for existence than groups that did not.

Williams constructs an alternative hypothesis, which makes no appeal to group advantage. Each organism evolves a behavioral strategy that maximizes its selfish advantage. When confronted by a threat, the organism will either flee or stand its ground. It is in the interest of an organism to flee when the attacker is large and threatening but to fight if the attacker is smaller and less dangerous. Williams postulates that the wolves that attack musk oxen appear large and threatening to females and young but seem smaller and less dangerous to adult males. This, he suggests, is why the individuals behave differently. Each acts in its own self-interest; the wagon-training behavior of the group is just a "statistical summation" of these individual adaptations.

After describing these two possible explanations of why musk oxen wagon-train, Williams says that we should favor the account given in terms of individual selection because it is more parsimonious. His point is not simply that the individual selection hypothesis is easier to understand or to test but that we should take the greater parsimony of the lower-level selection hypothesis as a *sign of truth*.

Perhaps there is something to the idea that group selection hypotheses are more complicated than hypotheses of individual selection. If a trait is common in some population and we interpret it as an example of altruism, then we need to think of a two-level process in which within-group selection and between-group selection work in opposite directions. On the other hand, if we interpret the trait as an example of selfishness, we need only describe a single selection process that occurs within the confines of a single population. However, even if we grant that individual selection hypotheses are more parsimonious in this sense, we must ask why parsimony should be relevant to deciding what we think is true.

Philosophers of science have thought a good deal about this issue. Popper (1959) argues that simpler hypotheses are more falsifiable; Quine (1966) contends that simpler hypotheses are more probable. These and other proposals are summarized in Hesse (1969). It is characteristic of these philosophical approaches that they seek a *global* justification for using simplicity as a guide to what we should believe.

My own view is that the justification for using parsimony (or simplicity—I use these interchangeably) to help decide what to believe depends on assumptions that are specific to the inference problem at hand (Sober 1988, 1990a). That is, I don't think there can be a *global* justification of Ockham's razor. My approach to this philosophical problem is a *local* one.

In the context of the units of selection problem, the basic question is whether the circumstances required for group selection to allow altruism to evolve occur commonly or rarely. If one believes that the kind of population structure needed for altruism to evolve rarely occurs in nature, then one should be skeptical of the suggestion that the behavior of musk oxen is an instance of altruism. However, if one thinks that the conditions are frequently satisfied or that they are probably exemplified in this case (though they may be rare in general), then it will make sense to take the group selection hypothesis seriously. The way to decide this question is not by invoking general methodological maxims but by doing biology.

There is no *a priori* reason to prefer lower-level selection hypotheses over higher-level ones. This preference is not a direct consequence of "logic" or "the scientific method" but depends on specifically biological hypotheses that should be made explicit so that their credentials can be evaluated. We will return to this question about the role of parsimony considerations in scientific inference in Section 6.6.

It is interesting to compare these various arguments against group selection with the view taken by Fisher (1930), Haldane (1932), and Wright (1945). None of them argued that group selection was a disreputable form of argumentation that no rigorous biologist should touch with a stick. Fisher took seriously the idea that sexual reproduction might owe its prevalence to the advantage it confers on the group. And Wright believed that his model of interdemic selection provided the population structure within which altruistic characteristics might evolve. Of course, none of these thinkers would have endorsed uncritical invocations of "the good of the species." For them, group adaptation was a scientific hypothesis that had to be judged on its biological merits. In contrast, much of the effect of Williams's (1966) book and of Dawkins's (1976) popularization has been to cast the concept of group adaptation into the outer darkness. Williams does grant that there is at least one well-documented case of group selection (to be discussed in the next section). And, as already noted, he does say that female-biased sex ratios would be *prima facie* evidence of group adaptation. Yet, the dominant tone of his book is that group adaptation is a kind of sloppy thinking. Dawkins's book expresses this attitude in its pure form.

In Chapter 2, I urged the importance of not confusing the properties of a proposition with the traits of the people who defend that proposition. Group selectionists have, at times, been uncritical of their pet hypotheses; however, this does not show that the hypotheses themselves are simply confusions that any clearheaded thinker can see are obviously mistaken. Perhaps the excesses of sloppy group selectionism elicited a gaggle of sloppy arguments against group selection. Be that as it may, there is no reason for us to accept either the naive endorsements or the fallacious criticisms of group adaptation.

4.5 Examples

In the previous section, I criticized a slew of arguments that attempt to show that the gene is the one and only unit of selection. Even if these arguments are rejected, the question remains of whether the selfish gene thesis is correct. The fact that an argument is flawed does not mean that its conclusion is false.

My own view is that the idea of the "selfish gene" is a good description of *some* traits, even though it isn't a good description of *all*. In this section, I'll describe a few biological examples that help establish the scope and limits of the three possibilities we have considered. The goal is to clarify how the concepts of genic adaptation, organismic adaptation, and group adaptation are related.

A very good example of a selfish gene is afforded by the process called *meiotic drive*. First, I'll explain what meiotic drive means when it occurs without complications; then I'll bring in some complications.

Individuals who are *Aa* heterozygotes normally produce half their gametes with *A* and half with *a*. This is a "fair" Mendelian process. However, there are alleles that garner more than their fair share of the gametes that come from heterozygotes. These are called *driving genes* or *segregator distorter genes*. Examples have been found in the house mouse *Mus musculus* and in *Drosophila*. Let us consider the consequences of meiotic drive when it has no effects on the survival and reproduction of the organisms in which it occurs.

Suppose the three genotypes have the same probabilities of reaching adulthood and produce the same number of offspring. In spite of this, the frequency (p) of the driving gene (D) and the frequency (q) of the normal gene (N) will change, as the following table shows:

	DD	*DN*	*NN*
adult frequencies	p^2	$2pq$	q^2
percent of *D* gametes	100%	$0.5 + d$	0%
frequency of *D* gametes	p^2	$pq + 2pqd$	0
frequency of *N* gametes	0	$pq - 2pqd$	q^2

Here, d measures the degree of distortion. Note that the frequency of the D gene is $p^2 + pq = p$ in adults but is $p^2 + pq + 2pqd = p + 2pqd$ in the gametes. This means that the frequency of the D gene will be greater in the next generation than it was in the previous one. When the process recurs in the offspring generation, the frequency of D goes up again. In the limit, D sweeps to fixation.

How often does the kind of process described in this model actually occur? The answer is that we do not know. If a distorter gene goes all the way to fixation, we no longer can find out that it *is* a distorter gene. To discover this, we must find the gene in heterozygotes and see what gamete frequency they produce.

Notice that the D gene evolves for reasons having nothing to do with its fitness effects on the organisms in which it occurs. As already mentioned, we can suppose that the organisms in the population have identical viabilities and fertilities. The driving gene evolves simply because it makes more copies of itself than does the gene against which it competes.

I noted earlier that Dawkins (1976, p. ix) thinks that "there are two ways of looking at natural selection, the gene's angle and that of the individual. If properly understood they are equivalent." The pure form of the process of meiotic drive shows that this is not correct. If genes evolved solely because they are good for the organism in which they occur, then D would not increase in frequency in the model just given. On the other hand, the driving gene is an excellent example of a selfish gene, properly so-called. Its evolution requires us to abandon a strictly organismic point of view.

When I introduced the idea of a driving gene, I said that examples have been documented in several species. But then I pointed out that in the model

Box 4.2 Junk DNA

Much of the genome seems to have no organismic function. Apparently, these large regions of "junk DNA" are noncoding; they play no role in the construction of the organism's phenotype. If we think of selection as evolving only those adaptations that benefit the organisms in which they occur, the existence of junk DNA should be very puzzling. From the organism's point of view, it is useless baggage; it imposes an energetic cost but provides no compensating benefit. However, if we take seriously the idea that genes can be selected *in spite of* their effects on organisms, the phenomenon is not so puzzling. Some genes are better than others at spreading copies of themselves throughout the genome. The existence of highly repetitive junk DNA may be the result of this genic process (Doolittle and Sapienza 1980; Orgel and Crick 1980). Just as in the case of meiotic drive, what is good for a gene may conflict with what is good for the organism.

just given, the driving gene will go to fixation, and so we will be unable to document the fact that the gene we see before us is, indeed, a driving gene. What gives? The answer is that the driving genes we know about in nature are more complicated than the model suggests. Besides affecting the segregation ratio, they also influence the viability and fertility of organisms. In these real cases, there is organismic selection against *D* because *DD* homozygotes are sterile or die before reaching reproductive age.

A more complicated model is needed to describe such real cases, one that describes the effects of two sorts of selection. First, at the level of gamete formation, selection favors *D* and works against *N*. However, at the organismic level, selection favors *N* and works against *D*. The result will be a compromise. *D* does not sweep to fixation, but neither is it driven from the scene. Rather, the population evolves to a stable polymorphism.

This two-part selection process involves two units of selection. *N* helps organisms, and *D* hurts them. But it also is true that *D* helps the chromosomes on which it sits, and *N* hurts the chromosomes on which it sits (by leaving them vulnerable to driving genes). In this case, a strictly organismic point of view *conflicts* with the selfish gene point of view. If adaptations evolve only when they are advantageous for the organisms possessing them, there should be no driving genes. Just as there can be conflicts of interest between the group and the organism, so there can be conflicts of interest between the organism and the gene.

I now turn to the one example that Williams (1966) concedes is a documented case of group selection in nature—Lewontin and Dunn's (1960) investigation of the *t*-allele in the house mouse. The interest of this example is not that Lewontin and Dunn got the biological details exactly right; rather, the point is to see what the idea of group selection amounts to and how an argument for its existence might be developed.

The *t*-allele is a driving gene, which renders homozygous males sterile. Lewontin and Dunn wrote a model like the one just sketched and deduced a prediction about what the frequency of the *t*-allele ought to be. They found that the observed frequency fell below the predicted value. To explain this observation, they had to postulate a third force influencing the gene's frequency; it was here that they appealed to the concept of group selection. House mice live in small local demes. If all the males in a deme are homozygous for the *t*-allele, they cannot reproduce, and so the deme goes extinct. Not only are their copies of *t* taken out of circulation; in addition, the females in the group, who also have copies of the *t*-allele, fail to reproduce as well.

In the process that Lewontin and Dunn postulate, three kinds of selection occur at once. There is gamete selection, selection against males who are homozygous, and selection against groups in which all males are homozygous for *t*. The first of these processes favors the *t*-allele; the second and third work against it.

In this example, organismic and group selection act in the same direction; both tend to reduce the frequency of *t*. I now want to describe a plausible example in which organismic and group selection oppose each other. This will provide an example of the evolution of an altruistic trait. The case I'll describe is the evolution of avirulence in the myxoma virus, discussed by Lewontin (1970). Again, the point is not that the biological details are still thought to be exactly the way Lewontin described them. Instead, the goal is to illustrate what the concepts of group selection and group adaptation really mean.

The myxoma virus was introduced into Australia to cut down on the rabbit population. After several years, two changes in the rabbits and in the virus were observed to have taken place. First, the rabbits increased their resistance to the disease. Second, the virus declined in virulence. The first of these changes has an obvious explanation in terms of individual selection. But how are we to explain the second? The virus is spread from rabbit to rabbit by a fly, which bites live rabbits only. This means that a highly virulent virus probably will kill its host before a fly comes along and spreads the virus to another rabbit. Less virulent strains keep their hosts alive longer and so have a better chance of spreading.

Lewontin (1970) says this process involved group selection. Why? The idea is that infected rabbits contain different strains of the virus, and more virulent strains replicate faster than less virulent ones. If this is right, then viruses of lower virulence are *altruistic*. They are less reproductively successful than viruses of higher virulence in the same group. However, groups of viruses that have lower virulence do better than groups with higher virulence. The two conditions that define the concept of altruism, depicted in Figure 4.3, are satisfied.

It is easy to lapse into a description of this example, according to which low virulence no longer sounds like an example of altruism. After all, high-virulence viruses quickly kill their hosts, while low-virulence viruses do not. Isn't it to the advantage of a virus not to kill its host? This makes low virulence sound like a kind of selfishness.

To see what is wrong with this reasoning, we must take to heart the distinction drawn in Section 4.3 between what is going on *within* each group and what is true *on average* in the ensemble of groups. Reduced virulence *evolves by natural selection*; this means that lower virulence is, on average, fitter than higher virulence. However, this says nothing about whether the trait is altruistic (unless one defines altruism as what cannot evolve). To decide this further question, we must consider *what occurs within groups (i.e., rabbits) in which strains of different virulence are present*. If low-virulence viruses replicate more slowly than high-virulence strains, lower virulence is a form of altruism.

4.6 Correlation, Cost, and Benefit

One of the most puzzling conceptual issues in the units of selection problem is the status of *kin selection* (Hamilton 1964). Many biologists insist that kin selection is not an example of group selection; for them, helping kin is conceptually on a par with parental care, and parental care is something that can be understood as a strictly organismic adaptation. Yet, there are other biologists who find it natural to view kin selection as a kind of group selection in which the interactors are relatives.

In this section, I will describe a very simple criterion for when altruism will evolve. It pertains to the case in which the individuals in a population pair up and then interact in some way that affects their fitnesses. We will see how this criterion for altruism to evolve is affected by adding the assumption that the paired individuals are related to each other (e.g., by being full siblings). Although this line of reasoning might be interpreted as showing that kin selection is really a special kind of group selection, this is not my main objective. The key to the evolution of altruism is *population structure*. It is important to understand how kin selection fits into that more general format.

A further benefit of this approach is that we also will be able to discuss the significance of game-theoretic approaches to the evolution of reciprocity (Maynard Smith 1982; Axelrod 1984). Reciprocity is a kind of conditional altruism. It is useful to be able to understand reciprocity within a more general context. And finally, it is worth supplementing the qualitative description provided in Section 4.3 for when altruism will evolve with something a bit more quantitative.

Assume that the individuals in a population are either altruists (A) or selfish (S). Altruists donate a benefit b to others and thereby incur a cost to themselves of c. Selfish individuals make no such donations. Individuals of either type will *receive* donations if they live with altruists. Altruism will evolve when altruists are, on average, fitter than selfish individuals. We will see that the criterion for $w(A) > w(S)$ depends on two quantities: the correlation between interactors and the cost/benefit ratio. The first quantity, it will emerge, depends on the rules followed in the population that determine who interacts with whom.

Suppose that a population is composed of n altruists and some number of selfish individuals. Suppose that each altruist donates a benefit b *to every other individual in the population*. In this case, the fitnesses of the two traits are:

$$w(A) = (x - c) + (n - 1)\,b \qquad\qquad w(S) = x + nb$$

In these expressions, x is the "baseline fitness." When altruists reduce their fitness by c units, this is a reduction from the fitness they would have had if they had not donated. In both expressions, the first addend describes the effect on the individual of its own phenotype, and the second describes the effect on the individual of the behaviors of others. Simple algebra shows that

(1) If everyone interacts with everyone, then $w(S) > w(A)$ if and only if $c + b > 0$.

This means that if donation confers a genuine benefit on the recipient ($b > 0$) and entails a genuine cost to the donor ($c > 0$), then altruism cannot evolve.

Now let's introduce some structure into the population. Let the individuals pair up; this may happen at random, or there may be a tendency for similar individuals to pair with each other. The paired individuals then interact in a way that affects their fitnesses. In this interaction, an individual's fitness is influenced both by its own phenotype and by the phenotype of the individual with which it has paired. The payoffs to the row player are as follow:

		You are paired with	
		A	S
You are	A	$x - c + b$	$x - c$
	S	$x + b$	x

When selfish individuals pair with each other, each receives the baseline fitness of x. When altruists pair with selfish individuals, the altruists suffer the cost (c) of donation, so their fitness is reduced to $x - c$; the selfish individuals, on the other hand, receive the benefit (b) donated by their associates but do not incur the cost of donation themselves. So, selfish individuals who are paired with altruists have a fitness of $x + b$. Lastly, we must consider the fitness that altruists have when they pair with each other. In this case, altruists pay the cost of donation but also receive a benefit from the donation of their associates; so altruists paired with altruists have a fitness of $x - c + b$.

Notice that whenever an altruist pairs with a selfish individual, the altruist always does worse than the selfish individual ($x - c < x + b$, on the assumption that $b + c > 0$). However, we now will see that this fact does not settle whether altruism is *overall* less fit than selfishness when individuals interact after forming into pairs. A population with this structure differs in a fundamental way from the unstructured everyone-interacts-with-everyone setup described by statement (1).

Given the payoffs just described, the fitnesses of A and S are as follows:

(2) $w(A) = (x - c + b)\,P(A/A) + (x - c)\,P(S/A)$
$w(S) = (x + b)\,P(A/S) + (x)P(S/S)$

Box 4.3 The Prisoners' Dilemma

A game called the (one-shot) Prisoners' Dilemma was first used by social scientists to characterize a problem about rational deliberation. Each of two players must decide whether to behave altruistically or selfishly during their single encounter. Each decides what to do independently of what the other decides. The 2 x 2 matrix given in the accompanying text describes the payoffs a player receives.

If $b,c > 0$, selfishness is the *dominant strategy;* this means that you do better by being selfish, no matter what the other player does. The same conclusion applies to your opponent. The result of rational deliberation is that both players choose to behave selfishly, so both end up worse off than they would have been if both had decided to be altruistic.

The Prisoners' Dilemma presents a somewhat pessimistic picture of what rational deliberation can produce. It shows how people can rationally deliberate with full information about the consequences and still end up worse off than they would have been if they had been irrational.

Analogously, the fitness function depicted in Figure 4.5 presents a pessimistic picture of what natural selection can produce. It shows how a fitter trait can displace a less fit trait, and yet, by the end of the process, the individuals are less fit than the individuals were before the process began.

Distinct from the one-shot Prisoners' Dilemma is the Iterated Prisoners' Dilemma, in which players pair and interact with each other some number of times. On each move, the players decide whether to act altruistically or selfishly. A strategy is a rule that tells a player what to do on each move: *Always be altruistic* and *always be selfish* are two unconditional strategies, but there are many strategies in which a player's move is conditional on the previous history of the game. In an Iterated Prisoners' Dilemma, which strategy it is rational to follow depends on the strategy followed by the other player. No dominance argument can be used to select a strategy.

Here, $P(S/A)$ is the probability that one individual in the pair is S, given that the other is A. Simple calculation shows that

(3) When individuals interact in pairs, $w(A) > w(S)$ if and only if $P(A/A) - P(A/S) > c/b$.

$P(A/A) - P(A/S)$ is a familiar statistical quantity; it is the *correlation* of the paired interactors (see Box 3.3 for the definition of correlation).

If groups form at random, then $P(A/A) = P(A/S) = P(A)$ and the inequality reduces to $0 > c/b$. This means that if there is genuine cost to donor and genuine benefit to recipient, altruism cannot evolve when pairs form at random. At the other extreme is the case in which like always associates with like. If $P(A/A) = 1$ and $P(A/S) = 0$, the criterion becomes $b > c$. In this case, altruism evolves precisely when the benefit to recipient exceeds the cost to donor.

When an altruistic behavior implies a particular cost/benefit ratio, statement (3) describes how much correlation there must be between interactors for the behavior to evolve. The more costly the donation (for a fixed benefit), the harder it is for the trait to evolve. And if the cost exceeds the benefit, c/b

will be greater than 1. Since correlations have unity as their maximum value, it is impossible for this sort of *hyperaltruism*, as we might call it, to evolve. Few of us would die to make someone smile. If our behavior were under the control of the kind of selection process described by statement (3) (a controversial assumption, to say the least; see Chapter 7), our reluctance to engage in hyperaltruism would be perfectly intelligible.

We've discovered so far that altruism may be able to evolve when populations are subdivided into interacting pairs in which the interactors tend to resemble each other. What could make altruists pair with altruists? One possibility is that relatives form into groups. This is the basic idea of Hamilton's (1964) theory of kin selection. When groups are composed of relatives, altruism can evolve because (or to the extent that) relatives resemble each other.

Statement (3) describes the general criterion for altruism to evolve when the population is structured into pairs. Now I'll explore the special case in which the pairs are full sibs. Hamilton's (1964) inequality states that altruism evolves precisely when $r > c/b$. The quantity r is the *coefficient of relatedness* of the interactors. It happens that (given some simplifying assumptions) full sibs are characterized by $r = 1/2$. We now will see that $P(A/A) - P(A/S) = 1/2$ if interactions are between full sibs, the population is randomly mating, and inheritance follows a symmetrical pattern that I'll now state.

Suppose that $A \times A$ parents produce 100 percent A offspring, $A \times S$ parents produce 50 percent A and 50 percent S offspring, and $S \times S$ parents produce 100 percent S offspring. This arrangement defines what is called the *haploid sexual* model of inheritance. If p is the frequency of A among the parents and mating is at random, then the three types of offspring sib groups occur with frequencies

$$P(AA) = p^2 + pq/2$$
$$P(AS) = pq$$
$$P(SS) = q^2 + pq/2.$$

This departure from Hardy-Weinberg frequencies is due to the fact that sibs tend to resemble each other. Note that the frequencies of A and S among the sibs do not differ from the parental frequencies; $P(A) = P(AA) + P(AS)/2 = p$ and $P(S) = P(SS) + P(AS)/2 = q$. These probabilities allow us to define the following conditional probabilities:

$$P(A/A) = P(AA)/P(A) = (p^2 + pq/2)/p = p + q/2$$
$$P(S/S) = P(SS)/P(S) = (q^2 + pq/2)/q = q + p/2.$$

Substituting these conditional probabilities into statement (2), we obtain

$$w(A) = (x + b - c)(p + q/2) + (x - c)(q/2)$$
$$w(S) = (x + b)(p/2) + x(q + p/2).$$

Simple algebra then entails that

Box 4.4 Kin Selection with a Dominant Gene for Altruism

Proposition (4) states a frequency *in*dependent criterion for the evolution of altruism. If c and b are constant, either altruism cannot increase when rare, or it goes all the way to fixation. Matters change if we vary the assumptions about inheritance. Instead of the symmetrical phenotypic rules already considered, let's suppose that altruism is coded by a single dominant gene.

Consider a diploid population in which individuals who are *aa* or *as* are altruistic (i.e., they have the A phenotype) and individuals who are *ss* are selfish (S). The allelic fitnesses then are

$$w(a) = P(a)w(aa) + P(s)w(as) = w(A)$$
$$w(s) = P(s)w(ss) + P(a)w(as) = P(s)w(S) + P(a)w(A).$$

This means that $w(a) > w(s)$ if and only if $w(A) > w(S)$. The altruistic gene (a) evolves precisely when the altruistic phenotype (A) has the higher fitness.

It follows that $w(a) > w(s)$ precisely when $P(A/A) - P(A/S) > c/b$. We now need to evaluate $P(A/A) - P(A/S)$ when interactions are between full sibs. The value of each term is frequency dependent:

$P(A/A) = 1$ when a is common.
$P(A/A) = 1/2$ when a is rare.
$P(A/S) = 3/4$ when a is common.
$P(A/S) = 0$ when a is rare.

$P(A/A) - P(A/S) = 1/2$ when the a gene is rare, but $P(A/A) - P(A/S) = 1/4$ when a is common. The criterion for altruism to evolve is frequency dependent.

This model has three possible solutions. If $c/b > 1/2$, selfishness goes to fixation; if $1/2 > c/b > 1/4$, the two traits evolve to a stable polymorphism; if $1/4 > c/b$, altruism goes to fixation.

(4) When interactions are exclusively between full sibs, $w(A) > w(S)$ if and only if $1/2 > c/b$.

Informal discussion of Hamilton's inequality $r > c/b$ for the case of full sibs often includes remarks like "full sibs share half their genes." Although this will be true in special cases, it is not true in general. In many populations, individuals are *very* similar to each other, and full sibs are even more so (Dawkins 1979). In addition, for altruism to evolve, it really doesn't matter how overall similar full sibs are to each other. What matters is the quantity $P(A/A) - P(A/S)$, where A and S are the two phenotypes (or the genes coding for them); the rest of the genome is quite irrelevant.

In the pairwise interactions considered so far, individuals do *not* pair at random (if altruism is to evolve), and they interact with each other *once*. I now will examine a different situation, one in which individuals *do* pair at random but interact with each other *repeatedly*. The pairs of individuals play an n-round Iterated Prisoners' Dilemma. On each move, they can either cooperate (be altruistic) or defect (be selfish). The payoffs on each move are as stated before. How well an individual does in this n-round game depends on the strategy he or she follows and on the strategy followed by the other individual in the pair.

One possible strategy that an individual might pursue is to act selfishly on every move. This unconditional strategy is called *ALLD* ("always defect"). Individuals may follow other, more complicated, strategies. Axelrod (1984) examines the strategy called TIT-FOR-TAT (*TFT*). An individual playing *TFT* will cooperate (i.e., be altruistic) on the first move and then will do on the next move whatever his or her partner did on the previous move. If the opponent cooperates, *TFT* does the same on the next move; if the opponent defects, *TFT* retaliates. *TFT* is a strategy that involves *reciprocity* (Trivers 1972).

Consider a population in which everyone follows either *TFT* or *ALLD*. Individuals pair and then play against their partners for n moves. The three sorts of pairs and the sequence of moves that occurs within each are as follows:

TFT *A A A* ...
TFT *A A A* ...

TFT *A S S* ...
ALLD *S S S* ...

ALLD *S S S* ...
ALLD *S S S* ...

Even if individuals pair at random, there still is an enormous amount of correlation between altruistic and selfish *behaviors* (Michod and Sanderson 1985; Wilson and Dugatkin 1991). The altruistic behavior encounters selfishness only during the first round of a game between someone playing *TFT* and someone playing *ALLD*.

We now can define the fitness of each strategy:

$$w(TFT) = n(x + b - c)\,P(TFT/TFT) + [x - c + (n - 1)x]P(ALLD/TFT)$$
$$w(ALLD) = nxP(ALLD/ALLD) + [x + b + (n - 1)x]\,P(TFT/ALLD).$$

If pairs form at random, $P(TFT/TFT) = P(TFT/ALLD) = p$ and $P(ALLD/ALLD) = P(ALLD/TFT) = q$. In this case $w(TFT) > w(ALLD)$ if and only if

$$n(x + b - c)p + [(x - c) + (n - 1)x]q > nxq + [x + b + (n - 1)\,x]p.$$

This simplifies to

(5) $w(TFT) > w(ALLD)$ if and only if $p(n - 1)\,(1 - c/b) > c/b$.

For fixed benefits and costs, whether *TFT* will be fitter than *ALLD* depends on the frequencies of the strategies and on the length of the game. Making *TFT* common (increasing p) and making the game longer (increasing n) both favor the evolution of *TFT*.

Consider an example. If $x = 1$, $c = 1$, and $b = 4$, the payoff matrix for each move of the n round game becomes:

		You are paired with	
		A	*S*
You are	*A*	4	0
	S	5	1

If there are fifteen rounds in each pairwise interaction ($n = 15$), statement (6) becomes $w(TFT) > w(ALLD)$ if and only if $p > 1/42$. When *TFT* is very rare, it cannot evolve, but once it crosses the threshold of $p = 1/42$, it goes all the way to fixation.

The payoffs to each strategy in the fifteen-round game may be derived from the previous payoff matrix, which describes the consequences of each move:

		You are paired with	
		TFT	*ALLD*
You are	*TFT*	60	14
	ALLD	19	15

Figure 4.5 describes the fitness functions of the two strategies. Note that there is an (unstable) equilibrium point at $p = 1/42$.

Let us now take stock. Paired individuals may be relatives or not; they may interact once or repeatedly. In all the circumstances just reviewed, the criterion for the evolution of altruism is the same. The degree of positive association between the interactors and the cost/benefit ratio determine whether altruism will evolve.

When an altruistic behavior is paired with a selfish behavior, the immediate effect is that the altruistic behavior does worse. However, by now it should be clear that this fact leaves open the question of whether the one behavior is fitter than the other. As in so many other problems, it is important not to mistake the part for the whole. When there are two traits in a population and the individuals pair up, there are *three* kinds of pairs. To be sure, when an altruist interacts with a selfish individual, the altruist does worse. But the fitness of altruism also reflects how well the trait does in groups in which both individuals are altruistic. Similarly, the fitness of selfishness involves not just its triumphs when paired with altruism but its self-defeating behavior when selfishness is paired with itself. The fitness of a trait is an average over how well it does in *all* the contexts in which it is exemplified. The clash of altruism and selfishness within a single group of two individuals is

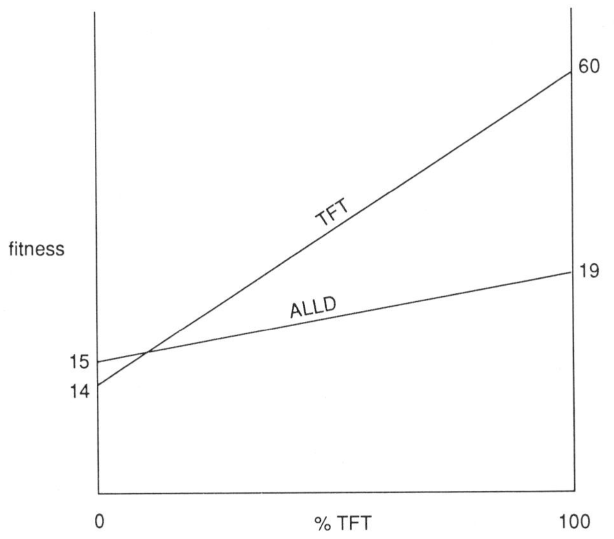

Figure 4.5 The depicted fitness relationships entail that the population will evolve to either 100 percent *TFT* or 100 percent *ALLD,* depending on the frequency at which the population begins. Each strategy is evolutionarily stable.

vivid, but that which is most striking sometimes fails to convey what happens *on average.*

Suggestions for Further Reading

A good sample of biological and philosophical work on the units of selection problem can be found in Brandon and Burian (1984). Wimsatt (1980) argues that genic selectionism is adequate as a form of "bookkeeping" for the results of evolution but does not adequately capture important features of the evolutionary process; Sober and Lewontin (1982) develop similar criticisms. Williams (1985) defends what he calls a reductionistic view of the units of selection problem. Sober (1981) connects the dispute over group selection with the problem in the social sciences concerning methodological holism and methodological individualism. Sober (1984b) describes a probabilistic model of causality, which he uses to characterize what the units of selection dispute is about. Lloyd (1988) defends Wimsatt's proposal. Brandon (1982, 1990) uses the probabilistic concept of "screening-off" to develop the point made by Mayr and Gould that selection acts "directly" on the organism's phenotype and only "indirectly" on its genes. Brandon's ideas are discussed by Mitchell (1987) and by Sober (1992b). Cassidy (1978), Waters (1986, 1991), and Sterelny and Kitcher (1988) develop a conventionalist approach to the units of selection problem, which is discussed in Sober (1990b). Wilson and Sober

(1989) argue that the contrast between group and organismic selection should be understood in parallel with the contrast between organismic and genic selection. Hull (1980) argues that a unit of selection, regardless of whether it is a group, an organism, or a gene, must be an "individual"; Sober (1992a) discusses this proposal. Hull (1988) generalizes Dawkins's (1976) notions of *replicator* and *vehicle* to provide a framework for understanding the units of selection problem.

5

Adaptationism

5.1 What Is Adaptationism?

Adaptationism is a thesis about the "power" of natural selection. Those who debate its truth do not doubt the tree of life hypothesis (Section 1.4). The dispute concerns the mechanism, not the fact, of evolution.

In order to understand what the debate is about, let's consider a very simple model of a selection process—the one I used in discussing the evolution of running speed in a population of zebras (Section 4.2). *Fast* competes against *Slow*, and the ultimate result is that the *Fast* trait goes to fixation. In this model, I pretended that zebras reproduce asexually and that an offspring always exactly resembles its parent. I also imagined that running speed evolves independently of all other characteristics—for example, that there is no correlation of *Fast* with a deleterious character (such as vulnerability to disease). I also supposed that mutation and drift had no effect on the evolutionary process.

All these assumptions are false. What would happen if we made the model more realistic by taking into account the complications just listed? Suppose we uncovered the genetic influences on speed and described how parental traits are transmitted to offspring in sexual reproduction. Suppose we explicitly recognized the fact that zebras don't live in infinite populations and that running speed may be correlated with other characteristics that matter to a zebra's fitness. And so on. Would these refinements affect our prediction about how running speed will evolve? Adaptationists will be inclined to answer this question in the negative. Their approach to the evolution of a trait holds that natural selection is such a powerful determiner of a population's evolution that complications of the kind just mentioned may safely be ignored. If the simple model predicts that *Fast* will go to 100 percent, adaptationists will expect a more complicated and realistic model to make basically the same prediction.

The same point can be made if we shift our attention to a more interesting selection model—Fisher's model of sex ratio evolution (Box 1.3). Fisher had no idea what the genetic mechanism is that allows a parent to influence the mix of sons and daughters she produces. He simply traced the fitness consequences of the different phenotypes he wished to consider. In the model, one assumes that the sex ratio trait evolves independently of other phenotypes (e.g., that there is no correlation between the sex *ratio* a parent produces and the *number* of offspring she has). And, of course, drift is ignored. Given these simplifying assumptions, Fisher's prediction was that the population will evolve to a point at which there is equal investment in daughters and sons. What would happen if we removed these idealizations? Adaptationists will expect that taking account of further complications isn't worth the candle. The prediction generated by the more complicated and realistic model will be the same, or nearly the same, as the one obtained from the simpler model in which selection is the only factor considered.

Adaptationists tend to expect nature to conform to the predictions of well-motivated models in which natural selection is the only factor described. They expect zebras to be *Fast* rather than *Slow* (if *Fast* is, indeed, the fitter phenotype). And they expect real organisms to invest equally in the two sexes (again, provided that this is the arrangement that natural selection is inclined to produce).

Of course, an adaptationist need not expect *all* populations to do what Fisher's model says they should. After all, Fisher's calculation is based on the assumption that there is random mating in the population. If the population is subdivided into sibgroups and if mating is strictly among sibs, then Fisher's model does not apply, but Hamilton's (1967) does. In this new setting, the prediction is that a female-biased sex ratio will evolve. Adaptationists well realize that which phenotype is fittest depends on the biological details.

Another qualification is needed in connection with the word "fittest." Adaptationists might expect zebras to evolve from *Slow* to *Fast* but will not expect them to evolve machine guns with which to counter lion attacks (Krebs and Davies 1981). When adaptationists say that the fittest trait will evolve, they mean the fittest of the traits *actually present in the population*, not the fittest of all the traits we can imagine.

Although adaptationists recognize that the outcome of selection is limited by the range of variation available, they often expect that range to be quite rich. As explained in Section 3.6, a correlation of an advantageous trait (like running fast) with a disadvantageous trait (like being vulnerable to disease) can prevent the fittest *combination* of traits from evolving. Adaptationists often view such correlations as temporary impediments, which the optimizing power of natural selection can be expected to overcome.

For example, consider the phenomenon of *antagonistic pleiotropy* (Section 3.6). Let A and a be two genes at a haploid locus. Each allele has two phenotypic consequences—one bad, one good. Suppose that phenotype P_1 would be better to have than P_2 and that phenotype Q_1 would be better to have than Q_2. Now let A cause both P_1 and Q_2 and a cause both P_2 and Q_1. Because the genetic system induces a correlation between a good value for P and a bad value for Q, the optimal arrangement—having both P_1 and Q_1—cannot evolve.

What would it take for natural selection to "overcome" this pleiotropic barrier to optimality? Dawkins (1982a, p. 35) says that "if a mutation has one beneficial effect and one harmful one, there is no reason why selection should not favor modifier genes that detach the two phenotypic effects, or that reduce the harmful one's effect while enhancing the beneficial one." A gene *m* at another locus might cause the *A* allele to produce both P_1 and Q_1. If so, that gene combination (*m* plus *A*) and the pair of advantageous phenotypes it generates would evolve to fixation. Dawkins's point is not just that this scenario is conceivable but that it is reasonable to expect that pleiotropic impediments to optimality are often overcome. Thus, natural selection will optimize with respect to *existing* variation, and it is reasonable to expect the existing variation to be *rich*.

Still, no adaptationist holds that variation is *limitlessly* rich. This is why zebras have not evolved machine guns and why pigs don't fly (Lewontin 1978; Dawkins 1982a). Take, for example, the case of antagonistic pleiotropy that Rose and Charlesworth (1981) discovered (Section 3.6). In female *Drosophila*, high fecundity early in life is correlated with low fecundity later. It may be biologically impossible for a mutation to allow fruit flies to have it both ways. In another context, Maynard Smith (1978b) pointed out that a zebra's running speed is increased by lengthening its leg, but lengthening its leg makes the leg more apt to break. The idea is that there is a mechanical constraint on leg design that prevents speed and strength from being optimized simultaneously. Adaptationists need not maintain that *all* impediments to optimality due to correlations of characters can be overcome. A more nuanced view is certainly possible; perhaps modifier genes can dissolve some correlations but not others.

How one characterizes the range of variation thought to be available in some ancestral population is a delicate matter. There is no precise view on this question that all adaptationists share. Adaptationism is a "tendency" of thought. In practice, its proponents often hold that variation is less constraining than critics of adaptationism are inclined to maintain. An extreme adaptationist will hold that *every* trait evolves independently of every other. An extreme antiadaptationist will hold that *every* trait is enmeshed in a web of correlations that makes it impossible to change a part without systematically changing the whole. Flesh-and-blood biologists rarely occupy either extreme. This does not mean that the contrast between adaptationism and its antithesis is unreal, only that there is no precise point on this continuum that separates adaptationism from its opposite.

We now can distinguish three theses about the relevance that natural selection has to explaining why the individuals in some population *X* possess some trait *T* (Orzack and Sober forthcoming):

(U) Natural selection played some role in the evolution of *T* in the lineage leading to *X*.

(I) Natural selection was an important cause of the evolution of *T* in the lineage leading to *X*.

(O) Natural selection was the only important cause of the evolution of *T* in the lineage leading to *X*.

These theses are presented in ascending order of logical strength; (I) entails (U) but not conversely, and (O) entails (I) but not conversely.

If (I) is true, then an explanation of the trait's evolution *cannot* omit natural selection; if (O) is true, then an explanation of the trait *can* safely ignore the nonselective factors that were in play. Adaptationism, as I understand the term, is committed to something like (O). For adaptationists, models that focus on selection and ignore the role of nonselective factors provide *sufficient explanations*.

Having described what it is to endorse adaptationism with respect to a single trait in a single lineage, I now can address the question of what adaptationism means *in general*. Adaptationists usually restrict their thesis to phenotypic characters. They often are prepared to concede that (O) and even (I) may be false with respect to molecular characters (Maynard Smith 1978b). This makes it reasonable to formulate adaptationism as follows:

> *Adaptationism*: Most phenotypic traits in most populations can be explained by a model in which selection is described and nonselective processes are ignored.

This is a generalization of (O).

Similar generalizations of (U) and (I) also are possible. The general form of (U) says that natural selection is *ubiquitous*. This claim is not terribly controversial. The generalization of (I) is a bit more substantial. (I) says that natural selection is an *important* cause of phenotypic evolution. If (I) were true in general, it would be a mistake to ignore natural selection. However, the debate about adaptationism centers not on this question but on the issue of whether, having taken account of natural selection, one can ignore everything else. The generalized form of (O) is the heart of the matter.

Adaptationism, as I construe it, does not demand that the process of natural selection maximize the fitness of the organisms (or the genes) in a population. As we saw in connection with the problem of altruism (Figure 4.3), natural selection can reduce fitness. Adaptationism emphasizes the importance of natural selection; it is not committed to the thesis that natural selection always improves the level of adaptedness.

Stronger versions of adaptationism can be obtained by replacing one or both occurrences of "most" with "all." The result of these substitutions would be to make adaptationism more falsifiable (Section 2.7). If adaptationism were the claim that natural selection suffices to explain *all* phenotypic traits in *all* populations, a single counterexample would be enough to refute it. However, few biologists would be prepared to endorse this strong form of the thesis. The formulation I am suggesting, though more difficult to test, is closer to the real issue that currently exercises biologists.

Box 5.1 The Two-Horn Rhinoceros Problem

Some disputes about adaptationism dissolve once the fact to be explained is clarified. Do rhinoceros horns require an adaptive explanation? That depends on what fact about the horns we wish to explain. Do we want to explain why *all rhinos have horns*? Or do we want to explain why *Indian rhinos have one horn while African rhinos have two* (Lewontin 1978)? The things that require explanation are *propositions*. An explanatory problem has not been specified properly until a proposition is formulated.

The two explanatory problems concern different patterns of variation. One might wish to explain why there is no variation in horn number *within* each of the two species, or why there is variation in horn number *between* the two species. Even if there is an adaptationist account of the former fact, it does not follow that there must be an adaptationist account of the latter.

5.2 How Genetics Can Get in the Way

If *Fast* is to supplant *Slow* in our population of zebras, *Fast* zebras must differ genetically from *Slow* ones. Phenotypic variation must reflect genetic variation. How confident should we be that, if a phenotypic trait comes under selection, there will be genetic variation to allow the trait to evolve?

It is sometimes suggested that we can reasonably expect that there has been ancestral genetic variation for any trait we care to name since artificial selection experiments normally succeed in changing the population. Although this is usually true, it is well to remember (1) that such experiments may involve a biased sample of phenotypic traits and (2) that sometimes the experiments fail. An interesting example of (2) is the efforts of animal breeders to change the sex ratio in dairy cattle. Selection has repeatedly failed to budge the ratio from unity; apparently, there is presently no genetic variation for sex ratio (Maynard Smith 1978a). It might appear that a 60:40 sex ratio isn't worlds apart from a 1:1 sex ratio, but in this instance, no existing genetic variant codes for the biased sex ratio.

Of course, the fact that there is presently no genetic variation for the trait does not refute Fisher's argument for why an even sex ratio evolved. After all, if the process that Fisher described is the correct explanation of the sex ratio we observe in dairy cattle, then we should expect that genetic variation for sex ratio was destroyed. Nonetheless, this example does have relevance to adaptationist assumptions about heritability. Adaptationists are inclined to maintain that if a phenotype were to become advantageous, then some gene combination coding for that alternative trait would probably arise (via mutation or recombination). Although this assumption is not always correct, the question of whether it is a reasonable working hypothesis remains open.

Adaptationists expect traits that have a significant influence on an organism's viability and fertility to be optimal. If *Fast* zebras are fitter than *Slow* ones, then present-day zebras should be *Fast*. This expectation may be mistaken if the population has recently experienced a major change in its environ-

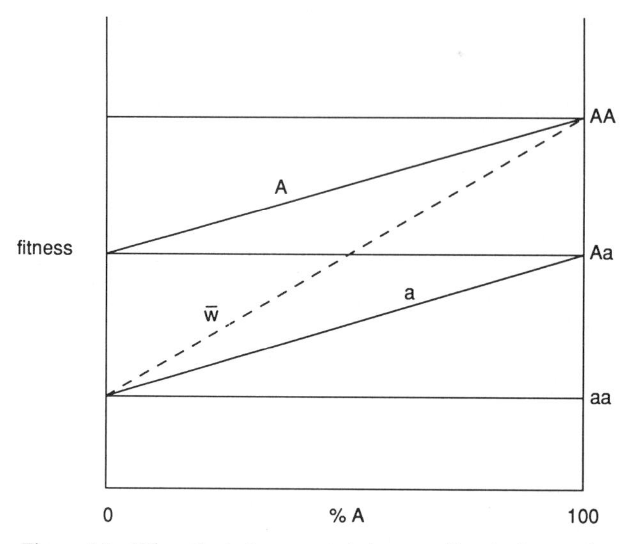

Figure 5.1 When the heterozygote is intermediate in fitness, the *A* allele evolves to fixation. At this point, all the individuals in the population possess the *AA* genotype, which is the fittest of the three genotypes.

ment. If this has occurred, then the population may not have had sufficient time for the optimal phenotype to evolve. In this case, the traits one presently observes will be *sub*optimal. Natural selection may be powerful, but even the most committed adaptationist will admit that it can lag behind extremely rapid ecological change (Maynard Smith 1982). It takes time for novel variants to arise and time for those traits to sweep to fixation.

In Chapter 4, I described how natural selection can make over the composition of a population (Figures 4.1–4.3). I considered purely phenotypic models in which inheritance follows the simple rule that like phenotype begets like phenotype. Nothing was said about how the different phenotypes are coded genetically. I now want to describe how genetic details can affect the power of natural selection. Even when mutation, migration, drift, and correlation of phenotypic characters are ignored, it still is possible for the genetic system to prevent the optimal phenotype from evolving to fixation. To see how this can happen, I'll consider three simple models in which natural selection acts on the genotypes at a single diploid locus. These models consider just two alleles, so there are three genotypes possible at the locus in question.

Figure 5.1 depicts the case in which the fitness of the heterozygote genotype falls between the fitnesses of the two homozygotes. Notice that the three genotypic fitnesses are frequency *in*dependent. From these genotypic fitnesses, we can calculate the fitnesses of the two alleles. When *A* is very rare, it almost always occurs in heterozygotes; when *A* is very common, it almost always occurs in *AA* homozygotes. So the fitness of *A* is frequency dependent,

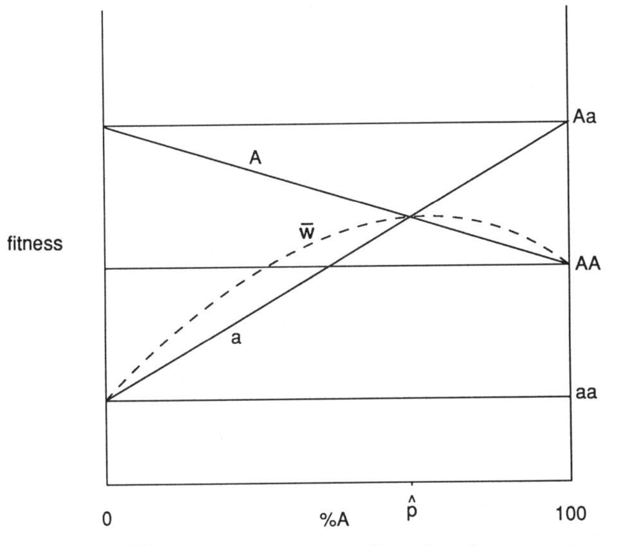

Figure 5.2 When heterozygotes are fitter than homozygotes, the two alleles are stably maintained in the population. In this case, the fittest genotype cannot go to fixation.

even though the fitnesses of the genotypes are not. The same point holds, of course, for the *a* allele.

Just to make the model more concrete, let's imagine that *aa*, *Aa*, and *AA* correspond to slow, medium, and fast running speeds in a zebra population. It is better to be fast than medium and better to be medium than slow. At every gene frequency, *A* is fitter than *a*. The result is that *A* goes to fixation; at the end of the process, all the individuals have the fittest genotype. All are *AA*.

A fundamentally different situation arises when the heterozygote is the fittest of the three genotypes (Figure 5.2). In this case also, the allelic fitnesses are frequency dependent. However, the result of selection favoring the fittest genotype is *not* that it goes to fixation. Rather, the population evolves to a stable polymorphism (at $\hat{p}$); the two alleles and all three genotypes are maintained. The reason the fittest genotype cannot go to fixation in this case is that heterozygotes do not "breed true." If the optimal phenotype is coded by a heterozygote, it cannot go to fixation. Here, the genetics of the system "gets in the way"; purely phenotypic considerations might lead one to expect that the fittest phenotype will evolve to fixation, but this cannot happen.

The last configuration to consider in a one-locus, two-allele model is heterozygote inferiority, depicted in Figure 5.3. In this case, the equilibrium frequency $\hat{p}$ is *unstable*. Although the alleles have the same fitnesses at this point, the population will not evolve *toward* this frequency but will go to either 100 percent *A* or 100 percent *a*, depending on where it begins.

With heterozygote inferiority, it needn't be true that the optimal genotype evolves to fixation. Where the population ends depends on where it begins. If

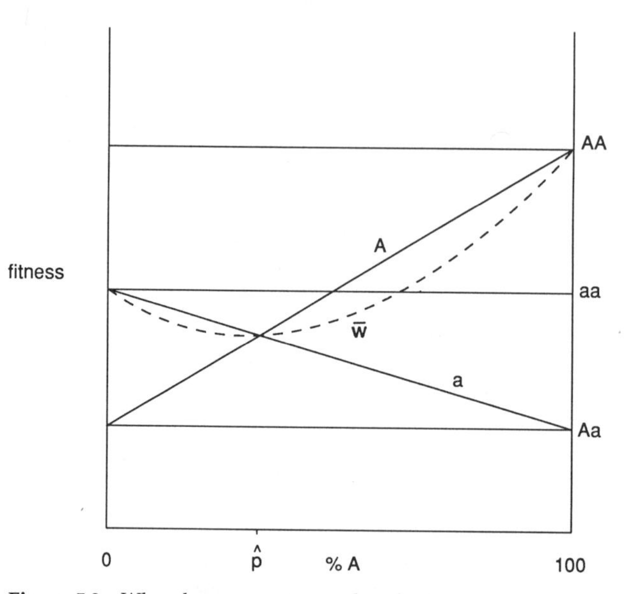

Figure 5.3 When heterozygotes are less fit than homozygotes, the population will evolve to either 100 percent *A* or to 100 percent *a*, depending on the frequency at which it begins. In this case, it is possible that the fittest genotype (*AA*) will fail to reach fixation.

a is the resident allele, the population cannot be "invaded" by a mutant *A* allele since *A* is less fit than *a* when *A* is rare. Although *AA* is the best genotype to have, fixation of *A* is not always "reachable" by the selection process modeled in Figure 5.3.

Of the three simple models just canvased, the last two would pose problems for adaptationism if they were widely applicable. If each genotype corresponds to a different phenotype, we can ask when the fittest phenotype will evolve. The fittest phenotype *cannot* evolve to fixation when it is coded by a heterozygote, and if the heterozygote is the least fit of the three genotypes, natural selection *may* fail to drive the fittest phenotype to fixation. Only when the heterozygote is intermediate is it true that the fittest phenotype *must* go to fixation (provided, of course, that no other force counteracts the effect of selection). So the truth of adaptationism depends on how often heterozygote superiority and heterozygote inferiority occur in nature.

It is a reasonable, though not exceptionless, rule of thumb that small changes in an organism's genotype produce small changes in its phenotype (Lewontin 1978). If an allele *A* provides "more" of some quantitative phenotype (like the ability to digest some nutrient) than the *a* allele does, then the heterozygote will be intermediate in its phenotypic value. However, it does not follow from this that the heterozygote will be intermediate in *fitness*. Perhaps heterozygotes have intermediate size, but it is an open question whether intermediate size is the best, the middle, or the worst of the three phenotypes an organism might have. Evolutionary theory provides no argument from first principles that shows that heterozygotes must have intermediate fitness.

Yet, population geneticists generally believe that there is enough empirical evidence to conclude that heterozygote superiority is a rare genetic arrangement (but see Maynard Smith 1989, p. 66). One of the few documented cases is the sickle-cell trait in our own species. Individuals who are homozygous for the sickle-cell allele suffer a severe anemia. Individuals who are homozygous for the other allele suffer no anemia but are vulnerable to malaria. Heterozygotes suffer no anemia and have enhanced resistance to malaria. When these populations live in malaria-infested areas, the sickle-cell allele is maintained in the population despite the deleterious consequences it has. It would be optimal to be both resistant to malaria *and* nonanemic, but this is a configuration that cannot evolve because of the way that genotypes code phenotypes.

When we turn to the question of heterozygote inferiority, we also must conclude that the phenomenon has been documented only rarely, but the reason for this is different. As noted earlier, if the heterozygote is inferior, one or the other of the two alleles will go to fixation. This means that once selection has run its course, we will not be able to see that the allele at fixation evolved by the process of heterozygote inferiority. It is arguable that if heterozygote superiority exists, we should be able to observe it; the same cannot be said for heterozygote inferiority.

The models depicted in Figures 5.1–5.3 describe how natural selection will modify a population when different genetic arrangements are in force. If adaptationism says that genetic details do not "get in the way," what does this mean? It does not mean that selection can produce evolution in the absence of a mechanism of inheritance. What it means is that fitter phenotypes increase in frequency, and less fit phenotypes decline. If each genotype in Figures 5.1–5.3 codes for a different phenotype, this simple rule of thumb tells us that the fittest genotype will go all the way to fixation. As already noted, this outcome is guaranteed in the first model but not in the other two.

Although the simple models just considered exhaust the possibilities for one locus with two alleles, there is much more to genetics than this. First, there can be more than two alleles at a locus. Second, an organism's fitness can be influenced by the combination of genes it has at more than one locus. In each of these cases, it is possible to describe some genetic arrangements that prevent natural selection from driving the fittest phenotype to fixation and others in which the genetics does not "get in the way." The debate about adaptationism involves the question of how common or rare these various arrangements are.

Adaptationists often talk about selection favoring a gene for this or that advantageous phenotype. This may give the impression that adaptationism is committed to the importance of studying specific genetic mechanisms. However, the fact of the matter is that adaptationism is a program based on the working hypothesis that phenotypic modeling is relatively autonomous. It maintains that the Mendelian details would not much alter the predictions made by purely phenotypic models.

5.3 Is Adaptationism Untestable?

I have devoted a fair amount of ink to enumerating some of the factors that can prevent natural selection from leading the fittest phenotype to evolve. Whether such factors are common or rare and how important they are when they arise is at the heart of the debate about adaptationism. In the light of all this, it is noteworthy that biologists have spent so much time arguing about whether adaptationism is testable. If my characterization of the debate is correct, the charge of untestability is really quite puzzling. For example, why should it be impossible to find out if antagonistic pleiotropy and heterozygote superiority are common or rare? And if we can answer these specific biological questions, won't that help settle the issue of whether adaptationism is true?

Adaptationism has been criticized for being "too easy." Suppose an adaptationist explanation is invented for some trait *T* in some population *X* and that we then find evidence against this explanation. The committed adaptationist can modify the discredited model or replace it with a different adaptationist account. Indeed, adaptationism seems to be so flexible a doctrine that it can be maintained no matter how many specific models are invented and refuted. The criticism lodged here is that adaptationism is *unfalsifiable* (Section 2.7): The complaint is not that adaptationism is a false scientific doctrine but that it is not a scientific claim at all.

Curiously enough, it is not just the critics of adaptationism who have asserted that empirical investigations do not test the hypothesis of adaptationism. Parker and Maynard Smith (1990, p. 27) say that when the optimality approach is used to address questions like "Why is the sex ratio often unity?" or "Why do dung flies copulate for 36 minutes?" then "the question is assumed to have an adaptive answer." In similar fashion, Krebs and Davies (1981, pp. 26–27) make it quite clear that one of their "main assumptions is that animals are well adapted to their environments." "We are not testing whether animals are adapted," they continue; "rather the question we shall ask ... is how does a particular behavior contribute to the animal's inclusive fitness." According to critics and defenders alike, adaptationism seems to be an assumption rather than a hypothesis under test.

What should we make of the claim that adaptationism is untestable? First, we must be careful to distinguish *propositions* from *persons* (Section 2.7). Perhaps some adaptation*ists* have been dogmatic; perhaps some have been unwilling to consider the possibility that nonadaptive explanations might be true. But this, by itself, says nothing about the testability of the propositions they hold dear. Whether adaptation*ism* is testable is a quite separate question from how adaptation*ists* behave.

The next thing to note about the thesis of adaptationism is that it does not admit of a "crucial experiment." There is no single observation that could refute the thesis if it is false. The word "most" that appears in the thesis is enough to ensure that there can be no crucial experiment. In addition, the thesis makes existence claims: It says that for most traits in most species, *there ex-*

ists a selective explanation. As noted in Section 2.7, existence claims are not falsifiable in Popper's sense.

The fact that adaptationism is not falsifiable in Popper's sense does not mean that it isn't a scientific statement. Rather, it means that there is more to science than is countenanced by Popper's philosophy. Adaptationism is like other *isms* in science. Like behaviorism and mentalism in psychology and functionalism in cultural anthropology, adaptationism is testable only in the long run. Its plausibility cannot be decided in advance of detailed investigations of different traits in different populations. Instead, biologists investigating a specific trait in a particular population are engaged in a process in which models are developed and tested against an ever-widening body of data. It is not absurd to think that, in the long run, we will arrive at biologically well-motivated explanations of various traits. If we can do this, we then will be able to survey this body of results and decide how often adaptationist explanations turned out to be correct. The idea that we must decide whether adaptationism is true *before* we begin the project of constructing and testing specific adaptationist explanations puts the cart before the horse.

Although no single observation will settle whether adaptationism is true, this does not mean that the thesis has no scientific importance. Generalizations about how evolution usually proceeds are of considerable scientific interest. Adaptationism is, so to speak, a "monistic" approach to the evolutionary process. An alternative to it is "pluralism," which holds that evolution is caused by a number of mechanisms of roughly coequal importance (Gould and Lewontin 1979). If adaptationism were beyond the reach of scientific investigation, pluralism would be untestable as well.

Adaptationism is first and foremost a *research program*. Its core claims will receive support if specific adaptationist hypotheses turn out to be well confirmed. If such explanations fail time after time, eventually scientists will begin to suspect that its core assumptions are defective. Phrenology waxed and waned according to the same dynamic (Section 2.1). Only time and hard work will tell whether adaptationism deserves the same fate (Mitchell and Valone 1990).

5.4 The Argument from Complex Traits

I have been arguing that adaptationism is a possible conclusion that one might draw about a specific trait in a specific population. It also can be generalized as a claim about phenotypic traits in general, across all of life's diversity. Whether one is an adaptationist about the sex ratio in a population of wasps or about all phenotypes in all populations, I have argued that adaptationism is not a premiss to use in one's investigations. However, it is a possible conclusion that one might draw after detailed biological models are developed and evaluated in the light of data.

I now want to discuss a very different outlook on adaptationism, one that seeks to show, in a somewhat more *a priori* fashion, that adaptationism is the right assumption to use when first examining the biology of any living system. This position is stated quite forcefully in Dawkins's (1982b) essay "Universal Darwinism."

Dawkins suggests that for a complex structure like the vertebrate eye, the hypothesis of natural selection is the only plausible explanation. If we rule out theological explanations and consider just the resources of contemporary evolutionary theory, selection is a plausible explanation; drift, migration, mutation pressure, and so on are not. According to Dawkins, the idea that the eye is adaptive is not a tenuous thesis that requires lots of detailed biological knowledge before it can be evaluated. Even before we invent selection models and test them in detail, we are entitled to be confident that natural selection is the right kind of explanation to seek.

There is much that I agree with in Dawkins's argument. However, I do not think it shows that "adaptationism," in the sense defined before, is correct. When Dawkins says that selection is far and away the most plausible explanation of such complex traits, what alternative explanations is he considering? It is true that drift *alone* is not a plausible explanation of the eye. The same is true for mutation pressure when it acts *alone*. If we had to choose a *single factor explanation* of the eye, that single factor would be natural selection. However, evolutionary theory allows us to formulate explanations in which more than one cause is represented. There are models that describe the combined effects of mutation and selection, of selection and drift, and so forth. The fact that natural selection is the best single-factor explanation does not tell us how to evaluate more complicated explanations.

In addition, it is important to realize that "the vertebrate eye" is a structure with a great many characteristics. To explain the evolution of this structure, one will have to assemble a great many explanations of why the structure possesses the characteristics it has. One feature of this structure is that it possesses a device (the cornea) that focuses incoming light. Another is that it has an iris that comes in a variety of colors. It is entirely conceivable that natural selection should be strongly implicated in explaining the first feature but not the second.

With these distinctions in mind, we can reformulate the issue about adaptationism with respect to the vertebrate eye as follows. As noted in Section 5.1, three claims might be made about the eye's features. One might claim that natural selection has played *some* role in the evolution of those features, that natural selection has played *an important* role in the evolution of those features, or that those features can be explained by models in which natural selection is the *only* consideration taken into account. These correspond to propositions (U), (I), and (O), respectively.

Perhaps it is plausible to accept the first and even the second of these propositions before one has amassed a great deal of biological information. This, I take it, is the point that Dawkins emphasizes when he claims that there is no serious alternative to the hypothesis that the eye is "adaptive." However, I would suggest that this is not enough to vindicate the thesis of adaptation*ism*. Even if natural selection has been important, is it also true that natural selection has virtually swamped the impact of all other evolutionary forces? This is by no means obvious but requires detailed model building and detailed data.

Notice that I do not construe adaptation*ism* as the thesis that the trait in question is an *adaptation* (Section 3.7). If the eye evolved because there was selection for having an eye, then the eye is an adaptation. But to say that this happened is not yet to say whether *other, non*selective processes also played a significant role. A cause does not have to be the only cause. Even if there was selection for eyes, perhaps the optimal structure was unable to evolve because of various constraints. Even if it is obvious *a priori* that natural selection is implicated, the issue of optimality remains open.

5.5 If Optimality Models Are Too Easy to Produce, Let's Make Them Harder

I have emphasized that there is nothing in the logic of adaptationism that prevents it from being tested. However, this does not mean that specific adaptationist hypotheses have always been tested with appropriate rigor. Nor does it mean that rigorous testing is always easy.

Rather than run through a rogue's gallery of deplorable examples, I want to describe a few cases of good adaptationist explanations. In saying that these models are "good," I do not mean that they are flawless or that they provide the final word on the traits they describe. Rather, what I wish to emphasize is that these models are not trivial. They made testable predictions that did not have to come true.

There is more than an ounce of truth in the charge that adaptationism is, at times, "too easy." This is especially true when the characteristic one wishes to explain is left vague. But adaptationists have found useful ways to sharpen the problems they want to address. In doing so, they have made it harder to invent hypotheses that fit the observations. A well-posed problem should not be too easy; adaptationists can claim to make progress when they pose and answer well-posed problems.

The first example to consider concerns size differences between the sexes. Suppose we focus on some single species—ourselves, for example—and ask why males are (on average) larger than females. Here are two hypotheses that might explain this observation: (1) Males and females exploited different food resources, and this division of labor led to size differences, and (2) males are larger than females because of sexual selection—females chose mates who were large (or, alternatively, larger males were more successful in their competition with other males for mates). With so much of human prehistory shrouded in obscurity, it may be difficult to know how to distinguish among these and other possibilities.

The problem becomes more tractable if we embed this one species into a larger context. Clutton-Brock and Harvey (1977) considered a number of primate groups that vary in the degree to which males are larger than females. They found no correlation between size dimorphism and niche separation. However, they did find that the *socionomic* sex ratio (the sex ratio in breeding groups, not to be confused with the species' overall sex ratio) is positively correlated with the degree to which males are larger than females. In monogamous groups, males and females are about equal in size. In polygynous

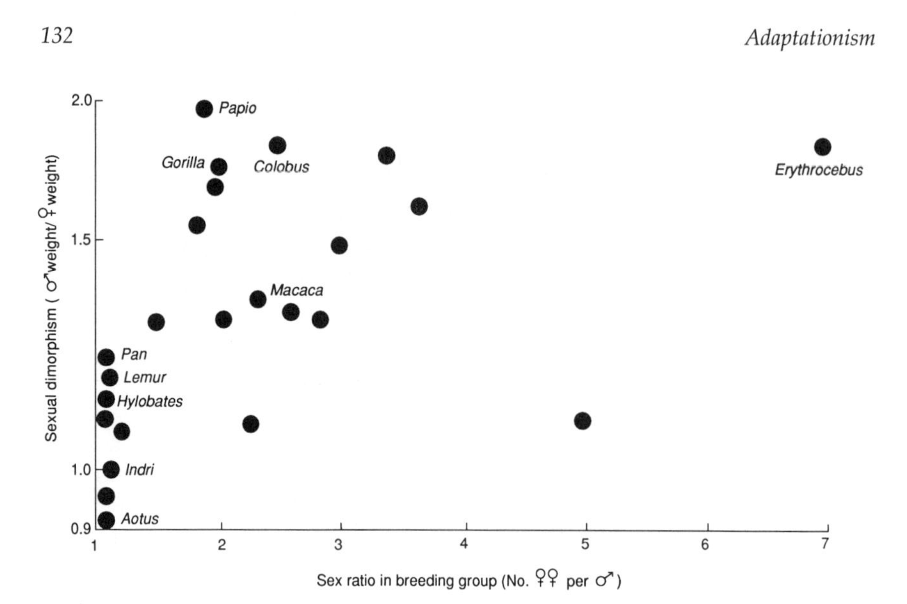

Figure 5.4 Clutton-Brock and Harvey (1977) found that there is a positive association in primate species between the degree to which males are larger than females and the degree to which females outnumber males in breeding groups.

groups, in which one or a few males breed with several females, males are much bigger than females. Clutton-Brock and Harvey took this finding to support the hypothesis of sexual selection since sexual selection would be more intense in species that are polygynous.

In saying that the observed correlation supports the hypothesis of sexual selection, several questions are left open. First, the data reveal a correlation, and correlation doesn't guarantee causation (Box 3.3). Perhaps some other explanation of the correlation, other than the hypothesis just mentioned, can be developed. If such an alternative can be formulated, it will pose a new testing problem, for which new data will be required.

Even if we tentatively accept the conclusion that sexual selection has causally contributed to sexual dimorphism in primates, we still haven't identified which other factors have contributed to this result. Figure 5.4 shows the data that Clutton-Brock and Harvey gathered. If we draw a smooth curve that comes as close as possible to these data points, this best-fitting regression line will have a positive slope. Note that the individual species will not all fall exactly on this curve. Why are some species above the line while others fall below? There is residual variance—variance that the hypothesis of sexual selection does not explain.

If the data points had shown very little dispersion around the regression line, we might have concluded that sexual selection is the only important cause of sexual dimorphism in body size. But, in fact, there is considerable dispersion, which suggests that sexual selection is, at best, one of the important causes of the character's distribution.

Where does this leave the question of whether adaptationism is the correct view to take regarding size differences between males and females in primates? It would be premature to declare victory or to concede defeat. After all, it is possible that natural selection influences sex differences in body size in ways that go beyond the impact of sexual selection on that trait. We cannot conclude from the data that (O) is false. This conclusion would be correct if a plausible *non*selective explanation were able to sop up the residual variance, but that remains to be seen. On the other hand, we also cannot conclude from the success of the sexual selection hypothesis that adaptationism is the right view of sexual dimorphism in primates.

Once again, we must be careful not to confuse the issue of whether there is an *adaptive* explanation of size dimorphism with the issue of whether size differences are *optimal*. Sexual selection is evidently an important part of the explanation of the data, so an adaptive explanation is relevant. However, this does not answer the question of whether any *non*selective factors also are important. This latter issue must be decided before we can conclude that the size differences found in different species are optimal.

This example illustrates the point that adaptationism is a conclusion we will be able to evaluate only in the long run. Clutton-Brock and Harvey studied a single characteristic in a single group—sexual dimorphism in body size in primates. We now have *some* understanding of this problem, but further research is needed, both to test existing hypotheses and to explain aspects of the variation that we do not now comprehend. In the long run, it may be possible to achieve a more complete grasp of this phenomenon. Then and only then will we be able to say whether (U), (I), or (O) is the correct view to take of this characteristic in primates. And, of course, this is just one trait in one group. Adaptationism is a thesis at a very high level of generality. Biology has a long way to go before it can say whether adaptationism is true.

The second example of an adaptationist explanation I want to examine concerns a single characteristic within a single species. It is Parker's (1978) investigation of copulation time in dung flies (*Scatophaga stercoraria*). When a fresh cowpat appears in a field, dung flies quickly colonize it. The males compete for mates. After copulating with a female, a male will spend time guarding her. After that, he flies off in search of new females.

What explains the male's guarding behavior? Females will mate with multiple males. Parker found that the second male fertilizes far more eggs than the first. He discovered this by irradiating males with cobalt; although irradiated sperm can fertilize eggs, the eggs do not develop. If an irradiated male copulates first and a normal male copulates second, about 80 percent of the eggs develop. If the mating order is reversed, only about 20 percent of the eggs develop. This shows that after copulation, a male has a reproductive interest in preventing the female from mating further.

The problem that Parker set for himself was to explain the amount of time that dung flies spend copulating. The observed value for this is 36 minutes, on average. Again by experiment, Parker found that increasing the copulation time increases the number of eggs fertilized. If copulation lasts for about 100 minutes, all the eggs are fertilized. However, there is a diminishing return on

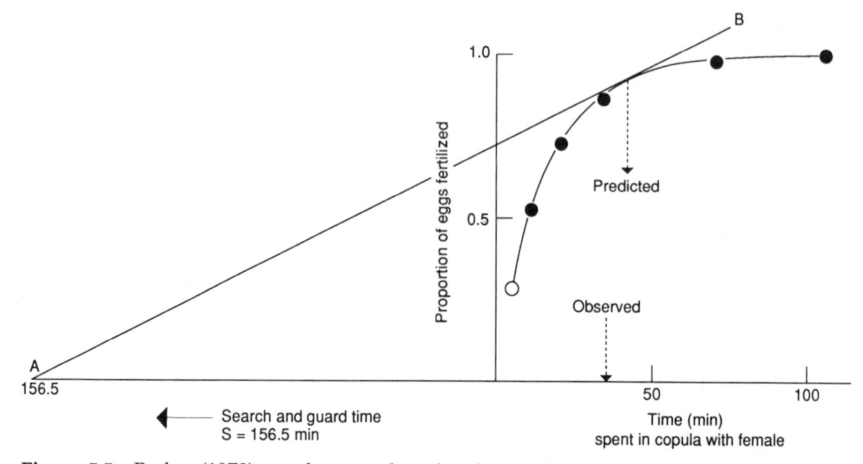

Figure 5.5 Parker (1978) sought to explain the observed copulation time in dung flies by calculating the amount of time that would maximize a male's rate of fertilizing eggs.

time invested—additional copulation time brings smaller and smaller increases in the number of eggs fertilized. The curve that Parker discovered is depicted in part of Figure 5.5.

Another factor to consider is that the time a male spends with one female is time not available for other copulations. Perhaps this suggests that males should copulate for some amount of time less than 100 minutes, after which they should seek a new mate. What copulation time would be optimal?

Parker measured the amount of time that males spend searching for new mates and guarding those mates after copulation. He found that the average for searching + guarding is 156 minutes. Since the total cycle for a male's reproductive behavior is search + copulate + guard, this total cycle will last 156 + c minutes (where c is the amount of time spent copulating). To find the optimal value for c, we need to find the value of c that maximizes the number of eggs fertilized *per unit time*.

Parker represented the problem graphically (Figure 5.5). The x axis represents the total time expenditure on the three tasks, and the y axis represents the number of eggs laid. For any choice of c, we can calculate the number of eggs that will be produced. To find the maximal *rate* of eggs fertilized, we need to find the triangle whose hypotenuse has the steepest slope. The upper tip of this hypotenuse must touch the curve that represents fertilization as a function of copulation time. Once we find that hypotenuse, we can derive from it the optimal value for c. The optimal value thus derived is 41 minutes, which is fairly close to the observed value of 35.

Parker's investigation exemplifies the motto that serves as this section's title: *If optimality explanations are too easy to invent, let's make the problem harder*. It is easy to cook up answers to vague questions about male dung flies. Why do they copulate? Answer: to maximize reproductive success. It is less straight-

Box 5.2 The Flagpole Problem

Hempel's (1965a) *deductive nomological* model of explanation says that we can explain why a proposition *P* is true if we can deduce *P* from a description (*I*) of initial conditions combined with a law *L*. Bromberger (1966) argued that these conditions do not suffice to explain *P*. Consider two examples.

A flagpole casts a shadow. We can deduce the length of the shadow from the height of the pole and the position of the sun, combined with laws that describe how light moves. This derivation of the length of the shadow *explains* why the shadow has the length it does.

Now suppose we wish to explain why the flagpole has the height it does. We can deduce the flagpole's height from the length of the shadow and the position of the sun, combined with laws that describe how light moves. However, this derivation of the flagpole's height does *not* explain why the pole has the height it does.

Why does the first derivation provide an explanation, while the second one does not? How does this issue pertain to Parker's explanation of dung fly copulation time? How is it related to the difference between cause and correlation (Box 3.3)?

forward to explain why they copulate *for 35 minutes rather than 5 minutes or 2 hours*. If a model answers a hard question, it will be less easy to make up another model that also fits the facts.

It would be easy but empty to say that the observed value of 35 minutes is the optimal solution to some unspecified problem. Parker's investigation went beyond this facile pronouncement in two crucial respects. First, he laid down a very specific *criterion of optimality:* The optimal copulation time is the one that maximizes the number of eggs fertilized per unit time. Second, he mustered independent empirical support for some of the crucial elements in his account. Rather than saying just that copulation time is subject to diminishing returns, Parker measured exactly what the return on investment is. The criterion of optimality, plus some measurements, allowed him to derive a specific prediction about what the optimal copulation time is. Here is a quantitative model that sticks its neck out.

There is one problem I want to raise about Parker's argument, however. Every adaptationist explanation assumes that the trait of interest evolved against some background of traits that already were fixed. Parker assumed that copulation time evolved in a population in which the search + guard time was already fixed at 156 minutes. Assuming that this is true, he derived a value for *c* that maximizes the rate of fertilization.

As far as I know, there is no reason to assume that the search + guard time was fixed at 156 minutes *before* copulation time evolved. Perhaps the temporal order was just the reverse, or perhaps the three time allocations evolved simultaneously.

If the three times coevolved, there might be different optimal trade-offs among them. Selection might constrain the sum of search + copulate + guard in some way but leave somewhat open what the component addends have to be. Selection then would be able to explain why the fly achieved an optimal set of time allocations *rather than one that is not optimal*. But selection would not be able to explain why one optimal set *rather than another optimal set* was the one that evolved. When the optimal solution to a design problem is not unique, neutral evolution may play an important role in explaining the characters we actually observe.

Male dung flies *now* spend about 35 minutes copulating, and they *now* spend about 156 minutes searching and guarding. If we wish to explain one of these traits, we must envision an ancestral population. But what characteristics should we assign to that population? If we assign to it all the *other* traits we currently observe, we are assuming that the trait we wish to explain evolved *last*. Obviously, this assumption must be backed up by evidence. So how do biologists *infer* a population's ancestral condition? Systematists call this the problem of *polarizing characters*—a problem I will discuss in Chapter 6.

In their review of optimization theory, Parker and Maynard Smith (1990) say that they wish to "lay to rest the idea that the application of optimization theory requires either that we assume, or that we attempt to prove, that organisms are optimal." The present example illustrates the point of this remark. Parker tested an optimality model by seeing if the value it predicted is close to the value observed in nature. If the values are close, one has evidence that the kind of selection described in the model is an important cause (not necessarily *the only cause*) of the character in question. Testing the model does not require one to assume that the general thesis of adaptationism is true or that the specific trait of interest is, in fact, optimal.

5.6 Game Theory

In many optimality models, the fitness of a trait does not depend on its own frequency. For example, in Parker's analysis of dung fly copulation time, his derivation of 41 minutes as the optimal copulation time did not turn on any assumption about whether this trait is common or rare. However, it is characteristic of models in evolutionary game theory that which behavior is optimal depends on what the other individuals in the population are doing.

A very simple example of this relativity was presented in the discussion in Section 4.6 of TIT-FOR-TAT (*TFT*) and *ALLD* (always defect). Figure 4.5 depicted the fitness relationship between these two traits. When *TFT* is common, that trait is fitter than *ALLD*. However, when *ALLD* is near fixation, *ALLD* is the fitter trait. Given these relationships (and the assumption that fitter traits increase in frequency), the population evolves to either 100 percent *ALLD* or to 100 percent *TFT*.

Each of these population configurations is *uninvadable*. If the population is 100 percent *TFT* and we introduce a few mutant ALLD players, the population will return to its initial configuration of 100 percent *TFT*. The same holds true for the configuration of 100 percent *ALLD*; if we add a few *TFT*ers, the population will return to 100 percent *ALLD*. A population configuration that is uninvadable is called an *evolutionarily stable state* (Maynard Smith 1982). Notice that uninvadability is relative to the alternatives considered; the fact that *ALLD* cannot invade a population of *TFT*ers does not address the question of whether some other trait might be able to do so.

An evolutionarily stable state is a property of a population. The more familiar concept in game theory—that of an *evolutionarily stable strategy* (ESS)—is different. A strategy is a policy that an individual follows that determines how it will behave. It is a property of an individual. A strategy *P* is an ESS if a population made of 100 percent *P* would be uninvadable.

It may appear that I am splitting hairs over the difference between state and strategy. After all, in the game of *TFT* versus *ALLD, TFT* is an ESS and 100 percent *TFT* is an evolutionarily stable state. However, to see that the distinction is a real one, consider the game of *Hawk* versus *Dove.*

Animals frequently engage in "ritualized combat." Rather than fighting to the death over some resource, conspecifics often engage in display or limited aggression, after which one of the contestants beats a hasty retreat. Lorenz (1966) and other ethologists had thought that this behavior evolved for the good of the species. Maynard Smith and Price (1973) investigated the *Hawk/Dove* game to show how restraint in combat can evolve by individual selection alone.

What might happen when two players come together to compete over some resource? Suppose the resource is worth 50 units of fitness, so that if two individuals were able to settle who got the resource without injury or loss of time, the winner would get 50 and the loser would get 0. Suppose further that the cost of serious injury is −100 and that time lost due to prolonged conflict entails a cost of −10.

We now need to consider what *Hawks* and *Doves* do in pairwise conflict. When a *Dove* encounters a *Dove,* they engage in a ritual tournament. Neither player gets hurt, and chance determines which player wins. The winner gets the prize (worth +50) but incurs a penalty of −10 for using up time. This gives the winning *Dove* in a *Dove* versus *Dove* match a score of +40. The losing *Dove* gets 0 with a time penalty of −10, yielding a score of −10. Since a *Dove* playing against another *Dove* has a 50/50 chance of winning, *Doves* get (0.5)(40) + (0.5)(−10) = 15 points, on average, when they compete with other *Doves.*

Next, let's consider what happens when a *Hawk* encounters a *Dove.* The *Hawk* quickly wins the prize, which gives it a payoff of 50. The *Dove* loses the resource, but it does so quickly and without serious injury. So its score in competitions against the *Hawk* is 0.

Lastly, we need to consider *Hawk* versus *Hawk.* A *Hawk* has a 50/50 chance of winning, in which case it gets +50 points. But there also is an even chance of serious injury, which costs −100. So the average payoff to a *Hawk* playing against another *Hawk* is (0.5) (50) + (0.5)(−100) = −25.

Here is a summary of your payoffs, conditional on who your opponent is:

		You play against a	
		Hawk	*Dove*
You are a	*Hawk*	−25	50
	Dove	0	15

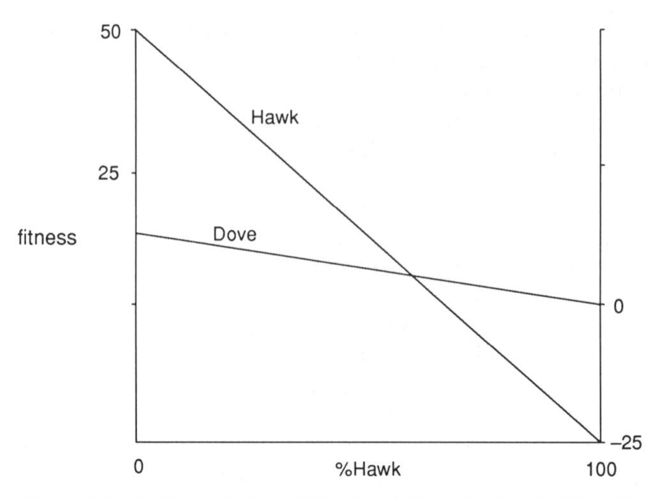

Figure 5.6 In the evolution of *Hawk* and *Dove,* the rare trait has the advantage. The result is that the population evolves to a frequency at which both strategies are represented.

How can we use this table to compute the average payoffs (fitnesses) of the two traits? Obviously, a *Hawk* does quite well when it plays against *Doves* but very poorly when it plays against other *Hawks.* The fitness of a *Hawk* will be a weighted average of these two payoffs, where the weighting reflects how often *Hawks* encounter one sort of opponent and how often they encounter the other.

The entries in this table represent the payoffs to the strategies in two extreme cases—when the population is (virtually) 100 percent *Hawk* and when the population is (virtually) 100 percent *Dove.* The payoffs at intermediate frequencies are not listed in the table but are represented in Figure 5.6.

We can infer from Figure 5.6 what the population's ultimate fate is. If the population begins with 100 percent *Doves* and a mutant *Hawk* is introduced, the *Hawk* does better than the residents, and so the *Hawk* trait increases in frequency. On the other hand, if the population contains 100 percent *Hawks* and a mutant *Dove* appears, the *Dove* will do better than the resident *Hawks,* and so *Dovishness* will become more common. Figure 5.6 says that advantage goes to rarity, so the stable equilibrium is a polymorphism. The population will evolve to a stable mix of 7/12ths *Hawks* and 5/12ths *Doves.*

In this example, a 7:5 mix of *Hawks* to *Doves* is an *evolutionarily stable state.* But notice that neither *strategy* is an ESS. It is false that a population made of 100 percent *Hawks* cannot be invaded by *Doves.* Ditto for 100 percent *Doves.* This illustrates the difference between describing a *population*'s evolutionarily stable state and saying that some strategy of an *individual* is an ESS.

A game is defined by the strategies considered and the payoffs they receive. There is no ESS in the game just described. However, if we modify the strategies, we can obtain a game very like that of *Hawk* versus *Dove* that does have an ESS. Suppose individuals can play *mixed strategies*. That is, an individual plays *Hawk* x percent of the time and *Dove* $(100 - x)$ percent of the time. Now we have a different game—instead of two pure strategies (*always* play *Hawk* and *always* play *Dove*)—we have an infinite number of mixed strategies (defined by different values for x).

In this new game, there *is* an ESS. It is the mixed strategy of playing *Hawk* 7/12ths of the time and *Dove* the other 5/12ths of the time. If everybody follows this strategy, the population cannot be invaded by a mutant following a different mixed strategy.

The two games just described have the same evolutionarily stable *state*, if we talk about *behaviors* rather than *individuals*. In each case, the stable equilibrium is a population configuration in which 7/12ths of the behaviors are *Hawkish* and 5/12ths are *Dovish*. In the first game involving the two pure strategies, the evolutionarily stable state cannot be achieved by an ESS. In the second game, it can be.

Models in game theory are usually applied to real-world examples by determining whether the observed configuration of the population is an evolutionarily stable state (or if some strategy that is observed to be at fixation is an ESS). When the observed configuration is an evolutionarily stable state, the conclusion is drawn that the model is a plausible explanation of why the population exhibits the configuration it does.

Several problems must be solved if such models are to explain real-world observations. I have already noted that selection models require some conception of the ancestral variation that selection acted upon. This will require an inference about the population's past, one that may be difficult if selection has destroyed past variation.

An additional difficulty is that the model builder must be able to measure the fitnesses of the various strategies described in the model. Even if one believes that a given population contains *Hawks* and *Doves* (and nothing else), numbers must be assigned to the payoff matrix. In practice, this is often difficult to do. For example, how is one to calculate the fitness cost incurred by prolonged display in a ritualized combat? Only when those values are provided can one calculate what the equilibrium frequency of the two traits should be.

It is tempting to downplay the importance of precise fitness estimates and to focus on the qualitative trends predicted by the model. In the *Hawk/Dove* game, the inequalities in the payoff matrix predict a stable polymorphism, quite apart from what the precise fitness values happen to be. Isn't this enough to explain the restraint in combat we observe in some natural population? The problem with this qualitative approach is that it makes it impossible to determine *how well* the model fits the data. An important strength of the optimality models described in the previous section is that they allow such quantitative questions to be posed and answered. If we wish to know how well adapted the organisms in a population are (relative to some model that tells us what they ought to be doing), the issue of quantitative fit of model to data cannot be evaded.

For example, the fitnesses specified before for *Hawk* and *Dove* predict that the population will evolve to an equilibrium of 7/12 *Hawks* and 5/12 *Doves.* But suppose we go to nature and find that the frequencies are, in fact, 11/12 and 1/12. Something has gone wrong; there are several possible diagnoses to consider. One is that the model is entirely correct so far as natural selection is concerned but some nonselective force has played an important role. Another possibility is that the model is entirely correct but that the population is still evolving toward the predicted equilibrium. A third possibility is that the fitness values in the model were incorrectly specified and that, once corrected, they would accurately predict the observed frequency. These and other options are very much worth exploring. However, such further questions would be invisible if we held that the model makes only the *qualitative* prediction that a *Hawk/Dove* polymorphism will evolve. Game-theoretic models are quantitative and deserve to be tested quantitatively.

When we talk of "testing" an optimality model, there are several propositions about the model that we might wish to investigate. One of them is that the selection process represented in the model has been an important influence on the trait under study. This is proposition (I) described before. A second possibility is that the optimality model takes account of all the factors that have had a major impact on the trait in question. This stronger claim is proposition (O). Whether we are talking about sexual dimorphism in primates, copulation time in male dung flies, or combative behavior in some species to which the *Hawk/Dove* model has been applied, it is important to keep these questions clearly separated.

I now want to consider another issue that is relevant to testing models in evolutionary game theory. It is important to remember that the ESS concept describes what the organisms in a population should be doing; it doesn't just describe some average condition of the population as a whole. This is something that game theory has in common with optimality models generally. If an optimality model predicts that the optimal wing length for some bird is 6.4 inches, we want to find out not just the population average but also the variance around that average. The organisms are optimally adapted if they each have wing lengths close to 6.4 inches. We cannot conclude that they are optimally adapted from the mere fact that the average wing has that length.

This point has important consequences when the ESS described in a model is a mixed strategy. Consider, for example, Brockman, Grafen, and Dawkins's (1979) study of nesting behavior in the digger wasp (*Sphex ichneumoneus*). A female digs a burrow, provisions it with live katydids that she has paralyzed, lays an egg on the katydids, and then seals the burrow. Alternatively, a female may enter an existing burrow, rather than digging one for herself. If the burrow she enters is empty (abandoned by its previous occupant), she proceeds. If it is already occupied by another female engaged in provisioning, the two wasps eventually meet (when both have returned to the burrow) and fight to determine which shall remain there. The average female goes through this cycle about twelve times in her six-week lifetime.

Brockman *et al.* modeled the evolution of the two behaviors, Entering and Digging. If most females Dig, Entering will be advantageous since there will be many empty burrows. However, if most females Enter, the existing burrows will be overcrowded, and so Digging will be advantageous. The fitnesses are frequency dependent, with the advantage going to the rare trait. Brockman *et al.* were able to calculate from their data the fitnesses of the two behaviors and to use these to predict what the evolutionarily stable state (the equilibrium population frequency) should be.

It is an interesting feature of this problem that the evolutionarily stable *state* can be realized in several ways. One is to have the wasps follow one or the other of two pure strategies—x percent always Dig, while $(100 - x)$ percent always Enter. Neither of these two pure strategies is an ESS. Another way the population can instantiate the equilibrium frequencies is by having all individuals follow the same mixed strategy of Digging x percent of the time and Entering the remaining $(100 - x)$ percent. In this case, the individuals in the population exhibit an ESS.

Of course, there is a third way that the population can realize the equilibrium frequencies of the two behaviors. Different individuals might follow different mixed strategies. Which of these three arrangements is the optimal one? At equilibrium, it makes no difference. At that frequency, the two behaviors have equal fitness, so an organism with one mix of Digging and Entering will have the same fitness as an organism with any other. But if the traits have undergone evolution, they weren't always at equilibrium. On the way to equilibrium, the ESS will always be favored by selection, but the same cannot be said of alternative strategies. So the model of this process doesn't just predict what the evolutionarily stable state will be; it also predicts that the population will evolve to a particular ESS. It predicts not just the mix of Digging and Entering found *in the population* but also that *each individual will exhibit that optimal mix*.

To test this prediction, Brockman *et al.* had to keep track of individual wasps. They couldn't just count how often the two behaviors occur in the population. Brockman handled this incredibly laborious undertaking. The wasps in one population she studied conformed to the ESS prediction, while those in another did not. The hypothesis of optimality turned out to be true in one case but not in the other.

This investigation is exemplary in the way it attends to individual variation. Unfortunately, ESS models are rarely tested in this way. Orzack and Sober (forthcoming) discuss this point in connection with sex ratio theory. Often, measurements are taken that indicate that the population average is close to the predicted value for the evolutionarily stable *state* and the conclusion is then drawn that the organisms are optimally adapted. Indeed, this point applies to optimality models generally. For example, Parker's study of dung flies (Section 5.5) reported *average* copulation times but gave no indication of how much variation there is among individual males. Optimality models in general and game theoretic models in particular describe what individuals ought to be doing, not merely what the average properties are that populations should exhibit. Tests must attend to individual differences.

I hope it is clear from my treatment of adaptationism that optimality models are important in evolutionary theory *whether or not adaptationism turns out to be correct*. Such models predict what characteristics organisms will possess if natural selection is the only important cause of character evolution. Biologically well-motivated models of this type are essential if one wishes to subject adaptationism to empirical test. Although critics of adaptationism often seem to think that optimality models are useless and irrelevant, this is the very opposite of the truth. Only by finding out what organisms *should* be doing if they are optimally adapted can one discover whether nature *actually* departs from the adaptationist paradigm.

Suggestions for Further Reading

The essays by Gould and Lewontin (1979) and Maynard Smith (1978b) helped set the parameters for subsequent debate about adaptationism. Chapter 8 of Oster and Wilson (1978) describes the components of an optimality model and points out errors that may arise in each. Krebs and Davies (1981) is a useful introduction to implementing the adaptationist research program in behavioral ecology. Maynard Smith (1982) systematizes the basics of the game-theoretic approach to evolution. Cain (1989) endorses a strong form of the adaptationist thesis, Beatty (1980) provides a philosophical treatment of optimality models, and Horan (1989) assesses the value of comparative data and optimality models in understanding adaptation. Mitchell and Valone (1990) discuss the issue of falsifiability in connection with Lakatos's (1978) ideas about progressive and degenerating research programs.

6

Systematics

Although Darwin called his most influential book *On the Origin of Species by Means of Natural Selection*, he often expressed doubts about the species concept. He says, "No clear line of demarcation has as yet been drawn between species and sub-species" (p. 51). He then notes that he looks "at the term species, as one arbitrarily given for the sake of convenience to a set of individuals closely resembling each other" (p. 52). Perhaps a less elegant but more apposite title for Darwin's book would have been *On the Unreality of Species as Shown by Natural Selection*.

If species are unreal, how could a theory aim to explain their origin? No book could explain the origin of centaurs, for example, because there are none. Of course, Darwin could and did shrug off this problem. His theory aimed to explain life's diversity. The upshot of the branching process he described is a multitude of organisms that are similar in some ways and different in others. Natural selection acts on differences among organisms and gives rise to differences among populations. The result of this process, Darwin thought, is that there will be no uniquely correct way to sort organisms into species.

Systematics is the branch of biology that seeks to identify species and to organize them into higher taxa, such as genera, families, orders, and kingdoms. It might seem that if species are unreal, the same will be true of the taxa to which species are said to belong. But curiously, this was not the conclusion that Darwin drew. The one diagram in the *Origin* shows how the process of descent with modification generates a *tree of life*. Darwin thought that this phylogenetic branching process provides *the* objective basis for taxonomy:

> All true classification is genealogical; that community of descent is the hidden bond which naturalists have been unconsciously seeking, and not some unknown plan of creation, or enunciation of general propositions, and the mere putting together and separating objects more or less alike (p. 420).

These passages from Darwin raise a number of interesting questions. What *is* a species? What distinguishes the higher taxa? In each case, we would like to know whether there is a uniquely correct definition or if there is a conventional element in the criterion one adopts.

An *extreme conventionalist* about a given taxonomic category (the species category, for example) will say that every grouping of organisms is just as entitled to be viewed as a species as every other. Few, if any, biologists would be inclined to adopt this extreme position. It would be absurd to place all green organisms or all organisms that weigh less than 35 pounds into a single species. But if we reject this extreme position, the question remains of how much freedom there is in our choice of a species concept. The diametric opposite of extreme conventionalism is *extreme realism,* which asserts that there is a uniquely correct choice of species concept. Between these two extremes are positions that involve different admixtures of realism and conventionalism. Where on this continuum is the most defensible position to be found?

In order not to short-circuit our inquiry into the reality of species and higher taxa, we need to distinguish the conventionalism at issue here from a quite different thesis, which I will call *trivial semantic conventionalism.* This kind of conventionalism holds that it is up to us what meanings we assign to the terminology we use. For example, there is nothing inherent in rocks that forces us to call them by the word "rock," rather than by the word "mush." This is true for *every* word we use, "species" and "genera" included. Semantic conventionalism is a universal and hence trivial thesis about how we pair words with meanings. Conventionalism is a philosophically interesting thesis only when it differs from trivial semantic conventionalism.

Conventionalism has been an important issue in the philosophy of physics in connection with the question of whether space is Euclidean. Newtonian theory endorses the Euclidean view, but general relativity says that space (or, more properly, space-time) is non-Euclidean (i.e., curved). The question is whether there is any evidence that decides between these two positions on the geometry of space. Geometric conventionalism says that there can be no evidence that indicates which geometry is true; we must choose a geometry on the basis of convenience. Note that this conventionalist thesis is quite different from the uninteresting claim of trivial semantic conventionalism.

Although this example from the philosophy of physics offers a useful paradigm of what philosophical conventionalism asserts, we cannot assume that the dispute between realism and conventionalism in systematics precisely replicates it. There appears to be a subtle but important difference between the two cases. It is generally agreed that the goal of both Euclidean and non-Euclidean geometry is to describe the geometry of space. The question is whether one of them does the job more adequately than the other. In contrast, disagreements in systematics are often at a more fundamental level: Systematists often disagree with each other about what the goal of classification is. As we will see, Darwin's thesis that "all true classification is genealogical" is controversial. Some reject it outright; others disagree in fundamental ways about what it means.

6.1 The Death of Essentialism

One reason science is of interest to philosophy is that it sometimes upsets and replaces received categories of thought. In our own century, relativity theory and quantum mechanics stand as admonitions to earlier generations of philosophers who had thought that various principles have the status of *a priori* truths. Kant not only held that space is Euclidean and that determinism is true; he thought that these principles are *necessary for the possibility of experience*. He maintained that a coherent and systematic physical theory cannot contradict these fundamental tenets. Since these principles are used to organize our experience, it follows that no experience could contradict them. The physics of the twentieth century showed that there is more to science than was dreamt of in Kant's philosophy. Far from being true and *a priori*, these principles turned out to be empirical and false. It is hard to imagine a more decisive overturning of the Kantian outlook.

Systematists in biology traditionally thought of themselves as describing the fundamental kinds of things that populate the living world. So entrenched was this outlook that the biological term "species" has long been a synonym for "kind." Yet, it has been argued that evolutionary theory shows that species are not kinds at all; rather, they are *individuals* (Ghiselin 1974; Hull 1978).

What is the difference between saying that species are natural kinds and saying that they are individuals? The fundamental distinction we need to draw is between a *property* and the *objects* that fall under that property. Redness should not be confused with the various fireplugs and apples that exemplify that property. A kind of thing is different from the things of that kind.

A kind may have zero, one, or many members. Furthermore, what makes those things members of the same kind is that they share the properties that define the kind. Consider the example of *gold*. This is a kind of substance. My wedding ring and the dome of the State House in Boston both fall under that kind because each is made of atoms that have atomic number 79. What is significant is that there is no requirement that my ring and the dome have a common ancestor or that they ever causally interacted with each other. Indeed, there may be gold in distant galaxies that has never causally interacted with the gold here on earth. Two individuals belong to the same natural kind in virtue of their *similarity*, not in virtue of their *history*.

Essentialism is a standard philosophical view about natural kinds. It holds that each natural kind can be defined in terms of properties that are possessed by all and only the members of that kind. All gold has atomic number 79, and only gold has that atomic number. It is true, as well, that all gold objects have mass, but *having mass* is not a property unique to gold. A natural kind is to be characterized by a property that is both necessary *and* sufficient for membership.

Essentialists regard the following generalization as describing the essence of gold:

> For every object x, x is a specimen of gold if and only if x is made of atoms that have atomic number 79.

To see why this generalization satisfies essentialist requirements, let us contrast it with another. Let's imagine that the universe (past, present, and future) contains only a finite number of gold objects and that we can list the spatiotemporal locations ($L_1, L_2, \ldots, L_n$) that each gold thing occupies. This allows us to formulate the following generalization about gold:

> For every object *x*, *x* is gold if and only if *x* is found at L_1 or L_2 or...or L_n.

Although both the displayed generalizations are true, the essentialist will regard only the first as specifying the essence of gold.

There are several reasons for this. First, the essence of gold must be given by a generalization that is *nonaccidental*. Gold things do not have to be found at precisely the locations listed, but they do have to have atomic number 79. Second, essences must be *explanatory*. Atomic number helps explain lots of the other properties that gold things have, but location does not explain much of anything. And lastly, an essentialist definition of gold must cite a property that is *intrinsic* to gold things; the cited property does not require that any relations obtain among gold things.

Essentialism is arguably a plausible doctrine about the chemical elements. Is it also a plausible view of biological species? That is, can we take a species (*Homo sapiens*, for example) and fill in the blank in the following skeleton?

> For every object *x*, *x* is a member of *H. sapiens* if and only if *x* is ______.

What is required is not just a true generalization but a necessary and explanatory one that specifies the intrinsic properties that make an object a member of *H. sapiens*. Is human nature to be conceptualized in the same way the essentialist conceptualizes the nature of gold?

We now can examine the idea that evolutionary theory refutes essentialism as a view about species. This claim has been defended in a variety of ways (Hull 1965). One argument goes like this:

> Natural kinds are immutable.
> Species evolve.
> ______________
> Hence, species are not natural kinds.

What does the first premiss mean? The idea is that although the members of a natural kind may change, the kind itself has a nature (an essence) that never changes. My wedding ring has changed in various ways, but the nature of gold has always been the same.

Once the first premiss is clarified in this way, we can see that the argument is flawed. Transmutation of the elements is possible; an atom smasher can transform (samples of) lead into (samples of) gold. However, this does not undermine the idea that the chemical elements have immutable essences.

Likewise, the fact that a population belonging to one species can give rise to a population belonging to another does not refute essentialism about species. Essentialists regard species as perennial categories that individual organisms occupy; evolution just means that an ancestor and its descendants sometimes fall into different categories.

Another argument against essentialism appeals to the fact that evolution is a gradual process. If species *A* gradually evolves into species *B,* where in this lineage should one draw the line that marks where *A* ends and *B* begins? Any line will be arbitrary. Essentialism, it is alleged, requires precise and nonarbitrary boundaries between natural kinds, and therefore, gradual evolution poses a problem for essentialism.

This argument, though less obviously flawed than the previous one, is not without its problems. To begin with, it presupposes a view of the speciation process that is no longer standard. Consider a single persisting lineage that begins with one suite of characters, several of which are gradually replaced by alternatives. At the end of the process, the lineage differs markedly from the way it was at the beginning. If the difference is great enough, one might be tempted to view the descendant population as belonging to a species that differs from the one to which its ancestors belonged. Such changes that occur within a single persisting lineage are examples of *anagenesis* (Figure 1.2).

Notice that the number of species that exist at a given time is not altered by anagenesis. One species existed at the beginning of the process and one species, though perhaps a different one, exists at the end. However, the idea of the tree of life requires that it be possible for the number of species to increase. This brings in a new idea—that speciation occurs when there is *branching* (*cladogenesis*).

Allopatric speciation occurs when a geographical barrier interposes itself between two populations that belong to the same species. If selection pressures differ in the two resulting populations, the characteristics of the two populations may diverge from each other. The divergence may be so dramatic that the two resulting populations will not interbreed even if the geographical barrier subsequently disappears.

Sympatric speciation also is a form of cladogenesis. However, in this case, the isolating mechanisms evolve without a geographical barrier first separating the two populations. An example is the process of *polyploidy*. A chromosomal accident may cause the offspring of a plant to have double (or triple or quadruple) the number of chromosomes that its parent possessed. If this happens to several members of the offspring generation, the polyploids may form a breeding population that is isolated from that of their parents.

Many systematists today reject the idea that speciation can occur anagenetically. A single species, like a single organism, may modify its characteristics while still remaining numerically the same species (Ghiselin 1974; Hull 1978). Increasingly, systematists embrace the view that *speciation requires cladogenesis* (an idea I'll discuss more fully in the next section).

This conclusion undercuts the objection to essentialism described earlier, which holds that essentialism fails because gradual evolution poses insuperable line-drawing problems. If speciation occurred anagenetically, separating one species from another might be impossible. But if speciation involves cladogenesis, it is arguable that a line can be drawn *sharply enough.*

Consider an example of allopatric speciation. A small group of rabbits is isolated from its parent population because a river changes course. Selection then leads this isolated population to diverge from the parental population; the isolated rabbits turn out to be the founders of a new species. We may wish to date the birth of the new species with the initial separation of the two populations or with the subsequent fixation of traits that prevent interbreeding between the two lineages. The point is that whichever proposal we follow, the cutoff point is precise enough. To be sure, there is no exact date (down to the smallest microsecond) for when that event occurred, but that degree of precision is not necessary.

The same conclusion can be drawn about cases of sympatric speciation. If polyploidy gives rise to a new species, one can give a satisfactory answer to the question of when the new species came into existence. Again, this does not require a date down to the smallest microsecond. Events, including speciation events, are spread out in time. The American Revolution was the beginning of a new nation; this event can be dated with *sufficient* but not *absolute* precision.

In general, essentialism is a doctrine that is compatible with certain sorts of vagueness. The essentialist holds that the essence of gold is its atomic number. Essentialism would not be thrown into doubt if there were stages in the process of transmuting lead into gold in which it is indeterminate whether the sample undergoing the process belongs to one element or to the other. I suspect that no scientific concept is *absolutely* precise; that is, for every concept, a situation can be described in which the concept's application is indeterminate. Essentialism can tolerate imprecisions of this sort.

To see why essentialism is a mistaken view of biological species, we must examine the practice of systematists themselves. With the exception of pheneticists (whose position will be discussed later), biologists do not think that species are defined in terms of phenotypic or genetic similarities. Tigers are striped and carnivorous, but a mutant tiger that lacked these traits would still be a tiger. Barring the occurrence of a speciation event, the descendants of tigers are tigers, regardless of the degree to which they resemble their parents. Symmetrically, if we discovered that other planets possess life forms that arose independently of life on earth, those alien organisms would be placed into new species, regardless of how closely they resembled terrestrial forms. Martian tigers would not be tigers, even if they were striped and carnivorous. Similarities and differences among organisms are *evidence* about whether they are conspecific, but a species is not *defined* by a set of traits. In short, biologists treat species as *historical entities* (Wiley 1981). They do not conceptualize species as natural kinds.

If biology had developed differently, the term "species" might now be used to label the various kinds that life forms can exemplify. The periodic table provides a list of chemical elements; the essence of each element is given by its atomic number. There is no *a priori* reason why there could not be a taxonomy of biological kinds in the essentialist's proprietary sense. Indeed, some evolutionists take seriously the idea that the diversity of life involves variations on the themes provided by a set of fundamental forms—*baupläne* (see, for example, Gould 1980a). The merits of this idea, rooted in a European tradition of ideal morphology, are currently being explored. However, the term "species" has been preempted for another role.

It is no accident that "species" has come to name historical entities rather than natural kinds. The list of chemical elements can be generated by enumerating the different possible atomic numbers. This list is not a heterogeneous and *ad hoc* hodgepodge; it is a principled consequence of atomic theory. The situation with respect to biological species is quite different. Neo-Darwinian evolutionary theory says that speciation is the result of a fortuitous confluence of biological and extrabiological circumstances. There is no theoretical principle that characterizes what the set of possible species must be. No wonder that the species concept was decoupled from its association with the idea of natural kinds.

The "fortuitous" character of evolutionary outcomes is a proposition specific to Darwinism; it is not inherent in the idea of evolution itself. Lamarck thought that each lineage moves through a preprogrammed sequence of forms: Life starts simple and then ascends a definite ladder of complexity. This is an evolutionary hypothesis in which major differences are not due to anything opportunistic. Within this Lamarckian framework, the idea that species are natural kinds could have found a happy home.

Even if *species* are not natural kinds, the idea of a basic system of natural kinds for life forms might be worth developing within a framework that is broadly Darwinian. If we visited another galaxy in which life had managed to evolve, what life forms would we expect to find there? If this galaxy is causally disconnected from earth, it will not contain any of the *species* found here. But perhaps we might expect to find flying, swimming, and walking creatures. Perhaps there will be creatures that extract energy from the nearest star and some that obtain energy by eating other creatures. What I have just said is very sketchy; moreover, it does not rest on any theory but merely reports what might be intuitively plausible. A *theory* of biological natural kinds, properly so called, would provide more details and would do so in a way that rests on principles rather than hunches.

This problem can be posed without thinking about life in other solar systems. If life on earth were destroyed and then evolved anew, what properties of present life forms would we expect to see repeated? Much of life's history is radically contingent; other characteristics may be more robust (Gould 1989). Models that elaborate this distinction will form an important part of theoretical biology.

6.2 Individuality and the Species Problem

Essentialism maintains that two things are both gold in virtue of their sharing some intrinsic and explanatory property that defines what gold is. Even if essentialism makes sense as a view about the chemical elements, it does not necessarily follow that we should adopt an essentialist interpretation of biological species. I have suggested, as an alternative to essentialism, that species are *historical entities;* this means that two organisms are conspecific in virtue of their historical connection to each other, not in virtue of their similarity.

In this section, I'll consider a further proposal about species—the idea that they are *individuals.* As we will see, individuals are historical entities, but not all historical entities are individuals. But before discussing whether *species* are individuals, I want to address the broader philosophical issue of individuality. What makes a car, a nation, a corporation, or an organism an individual? For complex objects such as these, the problem of individuation is usually approached by asking what it takes for two things to be parts of the same individual organism (or car, or nation, or ...).

One feature that all these examples have in common is that similarity is neither necessary nor sufficient for two parts to belong to the same individual. Let us focus on organisms to see why. Consider two identical twins, Jed and Fred. A cell from one of them and a cell from the other may be genetically and phenotypically as similar as you please, but they are not parts of the same organism. And two cells in Fred may differ from each other both phenotypically and genotypically (because of mutation, for example) while still being parts of the same individual.

If similarity is not the key, what is? For the case of individual organisms, the beginning of an answer may be found in the concept of causality. Two parts of the same organism causally interact with each other in characteristic ways. The cells in Fred's body influence each other; of course, it also may be true that the cells in Jed's body influence the cells in Fred's. However, the kinds of causal interaction at work in these two cases will be different.

The individuality of organisms involves a distinction between self and other—between inside and outside. This distinction is defined by characteristic causal relations. Parts of the same organism influence each other in ways that differ from the way that outside entities influence the organism's parts. The same point applies to things that are not organisms. For example, the parts of a car interact with each other in characteristic ways. And the parts of a nation or a corporation also have their characteristic modes of intra- and interindividual causal influence.

In the case of largish organisms like ourselves, the parts of an organism are spatially contiguous. If Fred and Jed are across the room from each other, then Fred's cells are belly to belly with each other, but they do not touch Jed's. Although this is a familiar arrangement for many organisms, it does not define the concept of individuality generally. The parts of a nation or a corporation need not be spatially contiguous. Alaska is part of the United States in virtue of the nexus of political interactions that unites the 50 states; the fact that Alaska does not spatially touch the lower 48 is immaterial (Ghiselin 1987).

The parts of an organism are united by relations of mutual biological dependence. The cells in your body help each other perform various tasks, and they are united by a common fate; an illness that impairs the function of one may well impair the function of others. However, once again, we must not mistake a peculiarity of organisms such as ourselves for some necessary characteristic of all living things. Organisms differ widely in the *degree of functional interdependence* that unites their parts. For example, the parts of a tiger are more functionally interconnected than the parts of an ivy plant. Excise an arbitrary 20 percent of a tiger and the tiger will probably die; excise an arbitrary 20 percent of an ivy plant and the plant may well live. Individuality is not a yes/no affair; it comes in degrees (Guyot 1987; Ereshefsky 1991b).

When a collection of parts shows little functional interdependence, doubts may arise as to whether the parts belong to a single organism. A stand of aspens connected by underground runners may be viewed as a single organism (Harper 1977; Janzen 1977; Dawkins 1982a). Yet, severing the runners does not much affect the viability of the parts. For this reason, we might be tempted to regard the stand as containing many organisms, not one. Rather than insisting on a definite solution to such problems, we perhaps should concede that the individuation of organisms has problem cases. As functional interdependence is reduced, a collection of parts becomes less and less of an individual. When do the parts belong to a single organism, and when are they each organisms unto themselves? There is no precise boundary.

So far, I have mainly concentrated on the *synchronic* problem of individuating organisms. Given that two parts exist during *the same period of time,* what makes them parts of the same organism? However, we also must consider the *diachronic* aspect of this issue. When two parts exist during *different periods of time*, when will they be parts of the same organism?

Once again, we can begin by discounting the importance of similarity considerations. Fred is now much different from the way he was as a child, but that does not undermine the claim that these two temporal stages are stages of the same enduring individual. Most of his earlier cells are gone, and some of the cells he now has are genetically different from the cells he had as a child. Indeed, it is possible to imagine that Fred the adult resembles Jed at age five more than he resembles Fred at age five. But here as well, we are not inclined to see Jed at age five as the earlier stage of Fred the adult.

As before, causal relations seem relevant to the problem. The influence of Fred at age five on Fred the adult differs from the influence that Jed at age five had on the way Fred is now. It is the causal processes of ontogeny, not facts about similarity, that unite an earlier and a later stage as two parts of the same enduring individual (Shoemaker 1984).

In the case of many sexual organisms, there is a rather precise divide between an organism and its offspring. The organism *grows* during its own lifetime; it also *reproduces*. Cell division (mitosis) underwrites the first process; gamete formation (meiosis) underwrites the second. Once an offspring exists, its physiological dependence on its parents declines. It is entirely natural to date the beginning of the new organism as the time at which egg and sperm unite. Of course, fertilization is a process spread out in time; only because we are so large and long lived are we tempted to describe the "moment" of fertilization as if it were instantaneous. Perhaps if we examined fertilization under the microscope, we would be unable to date the precise moment when the new organism begins.

The same can be said of the time at which an organism ceases to be alive. Death is not instantaneous; it is a process. The functional integration of an organism can deteriorate gradually. There are stages in the career of an organism in which it is clearly alive and stages in which it is clearly dead. In between, there may be gray areas.

These considerations concerning birth and death do not show that our ordinary concept of an organism is defective. True, the concept of an organism is not *absolutely* precise. However, for most purposes, the origin and demise of a sexual organism are well-defined *enough*.

It is an interesting fact about complex organisms that an organism can cease to exist even though its parts remain alive. Consider the following gruesome example. We dismember an organism, harvesting its organs for transplant purposes. We disperse its heart in one direction, its skin in another, etc. Each of these organs becomes part of a different recipient. The donor organism now no longer exists, not because its parts have ceased to exist but because it is no longer a functionally integrated whole. (The same point applies to our concept of a *car*.)

Further questions about how organisms should be individuated arise when we consider those that reproduce asexually. A hydra buds off a bit of itself, which then becomes a numerically distinct daughter hydra. The parent and the daughter are two organisms, not one. Though genetically identical, they are physiologically independent of each other. Let us call the parent Mom and the daughter Dee. It is natural to think of Mom as continuing to exist after she buds off Dee. Now consider two stages in Mom's career; there is Mom-before (= Mom before the budding-off event) and Mom-after. Why do we think of Mom-before and Mom-after as two stages in the career of a single organism? Why not say that Mom-before and Dee are parts of the same organism and that Mom-after counts as the new organism? Since the reproduction is clonal, there are no genetic differences among these three stages. And Mom-before is just as contiguous with Dee as she is with Mom-after.

The answer seems to involve considerations of *size*. When we face the problem of identifying Mom-before with either Mom-after or with Dee, we identify her with the larger piece. We can't identify Mom-before with *both* Mom-after and with Dee because Mom-after and Dee are distinct organisms. If we were to identify Mom-before with both, we would lapse into contradiction. We would have three objects b, a, d, and would find ourselves saying $b = a$, $b = d$, and $a \neq d$.

It follows from this way of thinking about the diachronic problem that binary fission poses an interesting difficulty. If Mom splits exactly in half and each half develops into a complete hydra, which half is the continuation of Mom? For the reason just stated, we cannot identify both halves with Mom-before. And since there is perfect symmetry between the left and the right halves, we cannot justify identifying Mom-before with one but not the other. So the natural decision is to identify her with neither. In the case of binary fission, the old organism ceases to exist, and a new pair of organisms comes into being.

A further complication arises when parts of the old organism are excised and immediately destroyed. When a piece of Mom is destroyed, Mom regenerates the missing piece and continues to exist. Indeed, if we destroy bigger and bigger bits, the old organism continues to exist if it regenerates the part that has been lost. It is a peculiar feature of our concept of an organism that whether we judge that the remaining bit is a continuation of the old organism or is an entirely new organism depends on what happens to the large bits that have been excised. If they are destroyed, the old organism continues to exist in the form of the small bit that survives. But if the large excised bit regenerates itself, then the smaller piece counts as a new organism entirely. (Hydras are like Thomas Hobbes's example of the ship of Theseus.)

I mention these issues to convey a feeling for just how intricate our ordinary concept of an enduring organism is. It is serviceable enough in most everyday contexts. However, I do not doubt that situations can be envisaged in which it is unclear how the concept should be deployed. The lesson, of course, is not that we should reject the concept. Rather, we must recognize that our concepts are not logically perfect. They, like organisms themselves, get along reasonably well in their normal habitats but may be seriously ill suited to coping with unusual circumstances.

I have argued that organisms are individuals because of several synchronic and diachronic facts. First, at a given time, the parts of an organism causally interact with each other in characteristic ways. Second, parts at different times are, for the most part, related to each other by a kind of ancestor/descendant relationship. The cells that now are in your heart are mostly descended from other cells that were in your body at an earlier time. (This, of course, does not mean that cells transplanted from another organism cannot *become* part of your body.) And finally, because of the dramatic difference between *growth* and *reproduction*, the lifetime of an organism has a reasonably precise beginning and end (Dawkins 1982a). These considerations, together, allow us to collect a set of parts and say that they are parts of a single, temporally enduring organism.

We now need to apply these ideas to the question of whether species are individuals. Advocates of this idea have argued that their position is a natural interpretation of Ernst Mayr's widely used *biological species concept*. Mayr (1963) held that a species is a group of actually or potentially interbreeding populations that are reproductively isolated from other such populations. He subsequently dropped the term "potentially" (e.g., Mayr and Ashlock 1991); it is this version of the biological species concept that the individuality thesis has endorsed. The organisms within a species reproduce with each other, but rarely, if ever, do they reproduce with organisms outside (Hull 1978; Ghiselin 1987).

To see whether species are individuals, I think it best to proceed in two steps. Initially, I'll consider the idea that a *breeding population* is an individual. Then, I'll consider whether a species can and should be thought of as a breeding population.

I'll use the term "breeding population" to denote a set of local demes linked to each other by reproductive ties but not so linked to demes outside the set. For example, the herd of deer living in a particular valley constitutes a single deme. Within the herd, there is reproduction. Moreover, this herd is reproductively linked to other such herds because of the entry and exit of individual organisms. A breeding population thus constitutes a *gene pool* whose parts are integrated by reproductive interactions.

The various earmarks of individuality noted earlier for organisms also apply to breeding populations. Causality, not similarity, is the key to their unity. A breeding population is born when it buds off from a parent population, just as a daughter hydra is born when it buds off from Mom. A new population comes into existence when daughter and parent become isolated from each other. Isolation doesn't mean that there is space between them but that gene flow is impeded. What makes parent and offspring two individuals rather than one is that they are, in fact, causally independent of each other. There is a physiological boundary between self and other in the case of organisms; there is a reproductive boundary between self and other in the case of breeding populations.

When does a breeding population cease to exist? If all the member organisms fail to reproduce and then die, that is enough for the whole to exit from the scene. But suppose the breeding unit holds together though its characteristics change? Here, the breeding population continues to exist, just as an organism does. Fred the adult has different traits from Fred at five, but they are two stages of the same individual. Breeding populations persist through time in the same way.

I noted earlier that an organism can cease to exist even though all its parts remain alive. This was the point of the organ transplant example. The same is true of a breeding population. Suppose the trout in Black Earth Creek breed with each other but do not, in fact, reproduce with trout in various other places in Wisconsin. If we take the trout from Black Earth Creek and disperse them to various other locales, the breeding population that was in Black Earth Creek has ceased to exist. Breeding populations are like nations and clubs in this respect: They can cease to exist as distinct entities even though the organisms in them continue to have babies.

One apparent point of difference between breeding populations and organisms is that the organisms in a breeding population are often less functionally interdependent than the cells in an organism are. The cells of a tiger depend on each other more than the trout in Black Earth Creek depend on each other. But surely this does not distinguish *all* organisms from *all* breeding populations. I have already mentioned that there are many organisms (e.g., ivy plants) whose parts are less functionally interdependent than are the parts of a tiger. Conversely, there are some breeding populations whose member organisms show a high degree of functional interdependence. Nests of social insects have sometimes been called *superorganisms* for just this reason (Chapter 4). Populations differ in their degree of individuality just as organisms do.

Let us consider with more care what it means to say that there is reproductive interaction inside a breeding population. It does not mean that every organism actually reproduces with every other during every instant of time. For example, in biparental species, the unit of reproductive interaction at any moment is a mating *pair*. Suppose that organisms *A* and *B* mate with each other at the same time that *C* and *D* mate with each other. Why do we say that all four of these organisms belong to the same breeding population? The point is twofold. First, the mating process was one in which *A could* have paired with *C*, and *B* could have paired with *D*. Second, let's suppose that descendants of the first pair will mate with descendants of the second. If so, the four organisms are part of the same breeding population in virtue of what will happen later (Sober 1984c; O'Hara forthcoming). A breeding population isn't internally integrated on a time scale of *microseconds* but on a time scale of *generations*.

Does this constitute a difference between the parts of organisms and the parts of breeding populations? Not fundamentally. When we say that the cells in an organism are causally integrated, this does not mean that each cell affects every other during every split second. Functional integration does not have to be *that* complete. The important point is that the potential for interaction exists and that interactions actually take place on a time scale appropriate for the kind of individual the organism is.

In population biology, organisms tend to be viewed as atoms. Their interactions with each other are highlighted, and their internal goings-on are shunted to the background. When a population biologist says that organisms are individuals, this does not deny that they are internally heterogeneous. From the point of view of another science, it may make sense to view an organism as a population of interacting cells. By the same token, the idea that breeding populations are individuals does not imply that they are internally unstructured. Whether a given collection counts as one individual or as many will depend on the magnifying power of the lens we use.

Having defended the idea that breeding populations are individuals, I now turn to the question of whether we should regard species as breeding populations. I won't try to answer this question fully; instead, I want to detail some of the costs and benefits entailed. We need to recognize the ways in which biological practice will have to change if this approach is adopted.

First and most straightforwardly, we will have to give up the idea that there are asexual species. The asexual organisms that exist at a time do not comprise a breeding population. They may trace back to a common ancestor that resembles them all, but the organisms at a given time go their own ways. A lineage of asexual organisms constitutes a historical entity, but it isn't a biological individual. Advocates of the individuality thesis have embraced the idea that there can be no asexual species (Hull 1978; Ghiselin 1987).

A second consequence is more controversial. Suppose a river separates some rabbits from their parent population and that the two populations remain phenotypically and genetically the same for as long as they exist (which, let us imagine, is a very long time). The individuality thesis says that these populations are not conspecific because they never *actually* interbreed. This conclusion is contrary to the practice of most biologists, who require distinct species to exhibit distinguishing characteristics.

How important is this problem? A defender of the individuality thesis might reply that if the two populations remain separated for long enough, their traits *will* diverge. If so, the strange consequence just described will not arise in practice. Alternatively, an advocate of the thesis might reply by biting the bullet—by saying that this is a surprising consequence of the proposal, which we should accept because the proposal is theoretically well motivated in other respects.

How often *do* situations of this type arise in nature? Ehrlich and Raven (1969) argue that many commonly recognized sexual species have subpopulations between which there is no genetic exchange. To evaluate this claim, we must be careful to consider the time scale being used. As noted earlier, the requirement of actual interbreeding does not demand that this happen during very brief intervals of time. Perhaps Ehrlich and Raven are right about what happens in the short term; if so, their point does not threaten the individuality thesis. That thesis conflicts with standard practice in this instance only if conspecific subpopulations are reproductively isolated from each other *on the appropriate time scale.*

Ehrlich and Raven (1969) and Van Valen (1976) also noted that there are many commonly recognized species that routinely form interspecific hybrids in the wild. Examples include species of North American oaks and of Hawaiian *Drosophila.* Is the individuality thesis committed to saying that these applications of the species concept should be rejected? I think not. Consider two populations, each with its own distinctive phenotypes and genotypes, that live in different locales. Suppose that between the two, there is a reasonably well-defined *hybrid zone;* hybrid organisms have characteristics that distinguish them from both their parents. I see no difficulty in saying that there are two breeding populations here. They happen to overlap, but each nonetheless retains its individuality. The individuality thesis could then conclude that there are two species here, not one.

The idea that distinct individuals may share parts with each other is not altogether alien. We count "Siamese twins" as two organisms, not one. And it is a familiar fact that individual nations can overlap in their areas of political control. After World War II, the United States, France, England, and the Soviet Union jointly controlled Berlin. By parity of reasoning, it may make sense to distinguish two breeding populations even when there is a hybrid zone between them. If so, the existence of hybrid zones is not, *per se,* an objection to the individuality thesis.

Another consequence of the individuality thesis is that anagenetic speciation is proscribed. A single lineage that changes its characteristics through time will count as a single breeding population, just as Fred at five and Fred the adult are two stages in the career of a single enduring organism. It isn't clear that this rejection of anagenetic speciation will make a tremendous difference in biological practice since our evidence about a lineage's distant past often comes to us in the form of fossils and we usually can't tell if a fossil is an ancestor of a present population rather than a near relative of it. Whether *Australopithecus* is our ancestor or our cousin isn't clear.

I so far have detailed some of the conceptual changes that the individuality thesis requires. The question we now need to face is whether these changes ought to be embraced. Do these implications constitute objections to the proposal, or are they surprising consequences that we should accommodate? This question requires us to examine what the alternatives are. If we reject the biological species concept, what species concept can we adopt in its stead?

The species problem is a conceptual issue in which three considerations are potentially at odds. A proposed solution should be evaluated for its *clarity*, its *theoretical motivation*, and its *conservatism*. It is easy to invent species concepts that are perfectly clear but that fall down on the other two counts. For example, if we grouped organisms according to their weight (species 1 includes organisms between 0 and 1 pound, species 2 includes organisms between 1 and 2 pounds, etc.), few line-drawing problems would arise. However, it is theoretically pointless to group organisms in this way. In addition, the proposal is totally orthogonal to biological practice. The proposal is an arbitrary stipulation, not a refinement of what biologists have been seeking to describe. (See Box 1.1.)

It also is possible for a species concept to receive high marks in the first two categories but fall down in the third. Consider some characteristic of organisms that is both reasonably clear in its application and biologically important. For example, some organisms always reproduce sexually, some always reproduce asexually, and some do both. This trichotomy allows us to divide the living world. Although the division is both (reasonably) clear and important, no one would propose that there are just three species. The reason is that this way of lumping and splitting has nothing to do with the use that biologists have made of the species concept.

I have pointed out various respects in which the thesis that species are individual breeding populations requires changes in practice. Although the proposal is somewhat conservative, it certainly does not leave everything as it was. I also have suggested that the proposal is reasonably clear and, moreover, that it focuses on biological considerations that are theoretically important. We now need to evaluate other proposals in the light of these three considerations.

As noted above, Ehrlich and Raven (1969) and Van Valen (1976) argue that commonly recognized species are sometimes not individuated by considerations of gene flow. As an alternative to Mayr's proposal, Ehrlich and Raven suggest that a species is made of organisms that are similar to each other in virtue of a common selection regime. And Van Valen (1976, p. 235) proposes an *ecological species concept:* "A species is a lineage ... which occupies an adaptive zone minimally different from that of any other lineage in its range and evolves separately from all lineages outside its range."

The main difficulty that these proposals must address is that it isn't clear when two populations confront "the same" selection regime or live in "the same" adaptive zone. Organisms within the same species live in somewhat different habitats and face somewhat different adaptive problems. If we recognize that some diversity in habitat is consistent with sameness of species, where do we draw the line? Symmetrically, there are many closely related species that occupy *similar* habitats. The difference between subspecific varieties, species, and higher taxa requires clarification.

A quite different objection to Mayr's species concept is developed by Sokal and Crovello (1970). They criticize Mayr's concept for not being operational. Biologists typically base their judgments about species membership on the phenotypic characteristics of organisms; scientists often do not observe whether the organisms they study successfully interbreed with each other. This is especially clear in the case of fossils. In addition, Sokal and Crovello think that the concept of reproductive isolation involves a "biased description" because it is theoretically loaded; they therefore reject Mayr's proposal and endorse a *phenetic species concept*. This is the idea that species are groups of organisms with a great deal of overall similarity.

One salient feature of this proposal is that the difference between species and taxa at lower and higher levels is left unclear. The organisms in a subspecific variety are more similar to each other than those in a species are, and the organisms in a genus will be less similar to each other than those in a single species are. The phenetic concept holds that there is nothing special about species as opposed to these other categories. In contrast, the biological species concept maintains that species have a unique status in the taxonomic hierarchy. Defenders of the biological species concept may view the difference between genera and families as a matter of convention. But species are different—they have a unique biological reality.

Questions also can be raised about how much the phenetic proposal departs from standard biological practice. Biologists routinely view males and females that breed together as conspecific, regardless of how little they resemble each other phenotypically. Consider, for example, a sexual species of lizard and an asexual species that is descended from it (Ereshefsky 1992). The asexual species consists of parthenogenic females. Perhaps from the point of view of overall similarity, we should group the sexual females together with the asexual females and treat both as distinct from the males. Almost no biologist would be willing to do this.

A further question that must to be posed about the phenetic species concept concerns the idea of overall similarity itself. But since this proposal about the species problem is part of a much larger view of systematics, I'll postpone discussing it until the next section.

I have mentioned only a few of the alternatives to the biological species concept that have been proposed, and these I have not explored in any detail. For this reason, I do not claim to have reached any definitive conclusion about which species concept to adopt. I have merely tried to lay out some of the pertinent philosophical issues.

Indeed, the idea that there is a single species concept that should be used in all biological contexts is not something we should assume dogmatically. Perhaps some form of *pluralism* is correct (Mishler and Donoghue 1982; Kitcher 1984). Pluralism should not be confused with *conventionalism*, according to which our choice of the species concept we adopt to describe a given biological situation is arbitrary. Pluralists maintain that we should use species concept *X* in some situations but concept *Y* in others. Conventionalists hold that whether we use concept *X* or *Y* in a given situation is arbitrary.

In conclusion, I want to return to the parallelism between the problem of individuating organisms and the problem of individuating species. In discussing hydras, aspens, tigers, and organ transplants, I described several principles that seem to govern the way we use the concept of a temporally enduring organism. I do not pretend to have provided a fully adequate treatment of this issue; identity through time is a difficult philosophical problem, and I have only scratched the surface. However, it is worth remembering that we feel quite reasonable in using the concept of an organism even though philosophers have yet to produce a fully adequate theory of the logic of that concept.

Many biologists have a quite different attitude to the species concept. The theory of how this concept should be used is quite unsettled. The question I want to raise is whether this should lead us to hesitate in using the species concept. Why should our talk of species have to await an adequate solution to the species problem if our talk of organisms need not await an adequate solution to the organism problem?

Is part of the answer that we ourselves are organisms and that our talk of organisms therefore strikes us as quite natural? Or perhaps the concept of an "organism" is part of common sense, whereas "species" is a theoretical concept in biology and so must live up to higher standards. A third possibility concerns how often problem cases actually arise in our experience. Although we may be puzzled whether a stand of aspens is one organism or many, we rarely encounter conundrums of this sort. In practice, we usually can apply the concept of an organism without difficulty, even though our theory of that concept is not fully developed. It is worth pondering to what extent the same point is true of our use of the species concept.

Although it is important to consider conceptual parallels between species and organisms, I don't want to deny a possible point of difference. This concerns the issue of functional interdependence of parts. If natural selection usually favors characteristics that are good for the individual organism but rarely favors traits that are good for the group (Chapter 4), it is no accident that we should find species more difficult to individuate than organisms. Natural selection will often have turned organisms into functionally integrated objects, whose parts interact in ways that benefit the whole. It will be no surprise if purely organismic selection often produces populations that show a much lower degree of individuality. Evolution may be responsible for the fact that the boundary between self and other is often clear in the case of organisms but can be more obscure when we consider species. If Darwin's theory deserves credit for solving the problem of the origin of species, then the process that theory describes may be blamed for giving rise to the conceptual difficulty we call *the species problem.*

6.3 Three Systematic Philosophies

Suppose we have before us a set of well-defined species. How should they be organized into a classification? In practice, this problem requires formulating principles that say how species should be arranged *hierarchically*. Nonhierarchical classifications are not impossible. Indeed, they were defended by some pre-Darwinian biologists (see, for example, William MacLeay's quinarian classification, discussed in Ospovat 1981), and outside biology, one example is alive and well in the *periodic* table of elements. How-

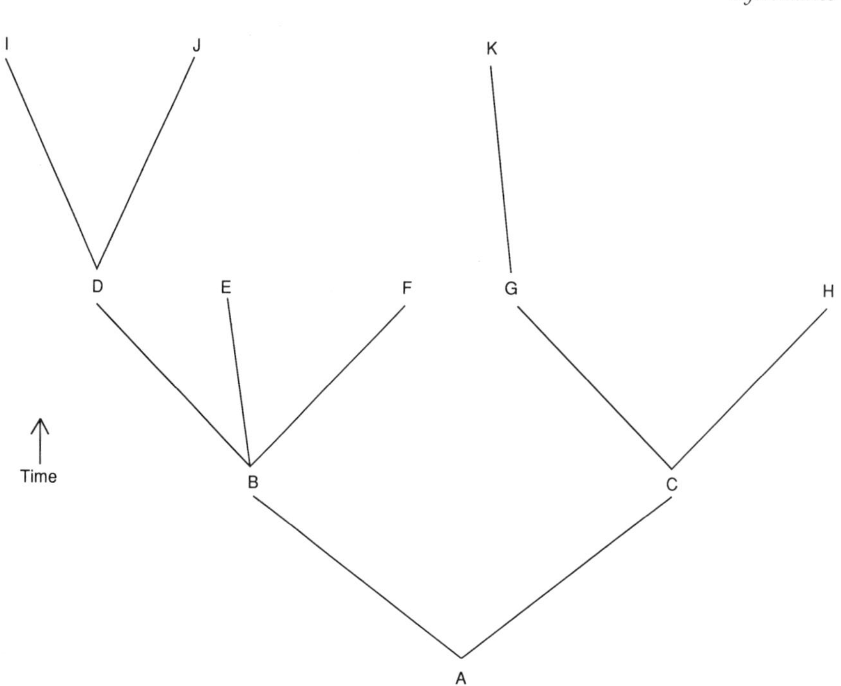

Figure 6.1 In a pure branching process, lineages split but never join.

ever, it is notable that, despite the foundational disputes now raging in systematics, the assumption of hierarchy is not at all controversial.

What is a hierarchical classification? A given species is a member of various higher taxa. For example, *Homo sapiens* belongs to *Mammalia,* to *Vertebrata,* and to *Animalia*. Each of these groups properly contains the one before. Another species—for example, the willow grouse *Lagopus lagopus*—belongs to *Aves,* to *Vertebrata,* and to *Animalia;* and these also are related as part to whole. Although *H. sapiens* and *L. lagopus* belong to some different taxa, there also are taxa that include them both. As we ascend the taxonomic hierarchy—from species to genus to family to order and so on up—a certain rule is always obeyed. Two species may belong to different lower-level taxa, but if they eventually get subsumed together in some higher-level taxon, they remain together at all subsequent higher levels. Since humans and willow grouse are both vertebrates, any taxon above the level of *Vertebrata* either must include them both or exclude them both.

In a hierarchical classification, there are only two possible relationships between a pair of taxa—proper containment and mutual exclusiveness. For example, *Aves* and *Mammalia* are mutually exclusive, but each is contained within *Vertebrata*. What is not permitted in a hierarchical arrangement is that two taxa should *overlap partially.*

Why should biological classification be hierarchical? If evolution is a

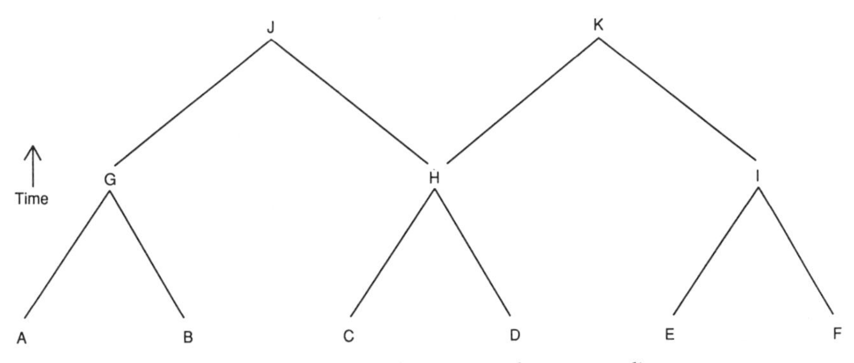

Figure 6.2 In a purely reticulate process, lineages join but never split.

branching process and if classification must strictly reflect that branching process, we would have an answer. However, both *if*s require clarification. Figure 6.1 illustrates what is meant by a *branching process*. An ancestor, *A*, gives rise to two daughters, *B* and *C*. These two objects, in turn, have the offspring depicted. Of these offspring, *D* has two daughters, *G* has one, and the rest have none. The objects represented might be species or they might be asexual organisms that reproduce by budding off progeny.

In this figure, the nodes represent the objects whose relationship is depicted; the branches represent the relationship of begetting. This interpretation differs from one that is often usual in systematics, according to which the branches are species and the nodes are speciation events. The reason I don't like the more usual interpretation is that it requires one to think that a parent species ceases to exist when it produces daughters. However, there is no reason to assume this: Just as an organism can continue to exist after it buds off a daughter organism, so a species can continue to exist after it buds off a daughter species.

What defines the idea of a branching process? For each object in Figure 6.1, there is a unique path back to the root of the tree. Or to put the point differently, each object has a unique immediate ancestor. As we move forward in time, branches split but never join.

This is not true for biological objects that reticulate rather than branch (Figure 6.2). Sexual organisms have *two* parents. If you wrote down your family tree, there would be more than one ancestor/descendant path from you back to previous generations. Likewise, when speciation occurs by hybridization, the resulting pattern involves joining as well as splitting.

How is the difference between branching and reticulate *structures* relevant to the issue of hierarchical *classification*? There is a simple way to define what counts as a taxonomic unit by using the idea of a branching structure. To do this, we first must define the concept of *monophyly:*

> A *monophyletic group* is a group composed of an ancestor and all of its descendants.

Box 6.1 Monophyly and the Species Problem

If the criterion of taxonomic reality is monophyly, can this cladistic idea be used to settle the species problem? Perhaps species, as well as superspecific taxa, should be monophyletic.

The problem with this proposal arises from the fact that species are sometimes ancestors of other species. Ancestral species *belong* to monophyletic groups, but they cannot *be* monophyletic groups. The cut method cannot isolate an ancestral species from the other species in a tree. Although the requirement of monophyly makes excellent sense for superspecific taxa, it is not an appropriate requirement to place on the species concept itself. This consideration counts against the *phylogenetic species concept* (Mishler and Donoghue 1982).

This concept can be applied to the branching structure depicted in Figure 6.1 by what I call *the cut method*. Draw a cut across any branch. The nodes immediately above that cut comprise a monophyletic group: *B, D, E, F, I,* and *J* comprise a monophyletic group, and so do *D, I,* and *J*.

An important fact about monophyly is that the complement of a monophyletic group is not itself a monophyletic group; if we "subtract" *D, I,* and *J* from the taxa depicted, the remaining species do not constitute a group made of an ancestor and *all* its descendants. Another consequence of the concept is that the monophyletic groups in a branching structure may have one of two relationships to each other but not a third. Either they are mutually exclusive or one is contained in the other; they cannot overlap partially. In short, *the set of all monophyletic groups defined on a branching structure constitutes a hierarchial classification.*

The cut method also defines monophyletic groups in a reticulate structure like the one depicted in Figure 6.2. However, in this case, such groups *can* overlap partially. The cut method says that G and *J* comprise a monophyletic group and that *H* and *J* do so as well. The requirement of monophyly does not generate a hierarchial classification in such cases.

This does not mean that there cannot be hierarchial classifications for species that come into existence by hybridizing. What it means is that the requirement of monophyly does not *suffice* to generate such a classification. Some other organizing principle is needed.

I have just described a sufficient condition for a classification's being hierarchical. If the objects considered form a branching structure and if taxa are required to be monophyletic, then the classification will be hierarchical. This is the cladistic taxonomic philosophy, stemming from the work of Willi Hennig (1965, 1966). I now will explain the other two taxonomic philosophies that have been taken seriously by biologists. These see hierarchical classification as the goal of systematics but seek to base that hierarchy on quite different conceptualizations.

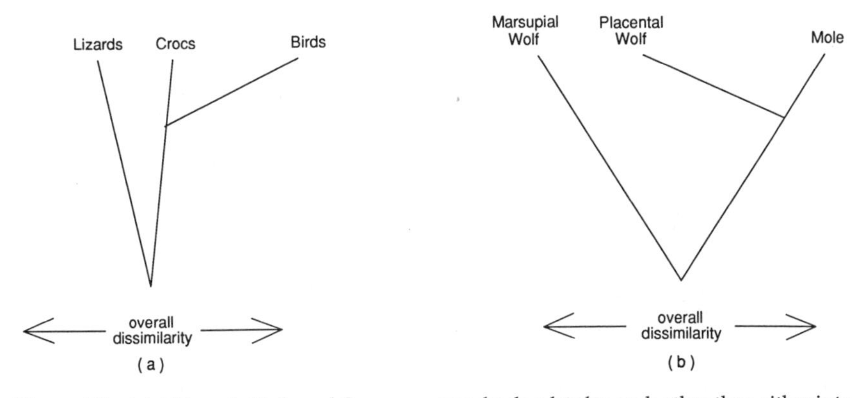

Figure 6.3 (a) Although Birds and Crocs are *more closely related* to each other than either is to Lizards, Lizards and Crocs are *more similar* to each other than either is to Birds. (b) Although Moles and Placental Wolves are *more closely related* to each other than either is to Marsupial Wolves, Marsupial and Placental Wolves are *more similar* to each other than either is to Moles.

Pheneticism defines taxa by the *overall similarity* of their members. Organisms are grouped into species by a criterion of resemblance, then the species are formed into genera by the same process and so on. Pheneticism's bottom/up approach to similarity grouping guarantees that the resulting classification will be hierarchical.

Before considering the adequacy of phenetics in the next section, we need to see how phenetic and cladistic principles can come into conflict. Overall similarity does not always reflect relations of monophyly.

There are two circumstances in which these concepts can disagree, which are illustrated in Figure 6.3. In pattern (a), birds diverged from their ancestors, while crocodiles and lizards were highly conservative, retaining many of the features of their common ancestor. The result is that lizards and crocodiles are more similar to each other than either is to birds. Nonetheless, crocodiles and birds have a common ancestor that is not an ancestor of the lizards. The phenetic grouping would be (Lizard,Crocodile)Bird; the cladistic grouping would be (Crocodiles,Birds)Lizards. The cut method isolates Crocodiles + Birds, not Lizards + Crocodiles, as a monophyletic group.

In pattern (b), placental wolves and marsupial wolves each independently evolved a similar cluster of novelties. In terms of overall similarity, they belong together. But in terms of genealogical relatedness, it is placental wolves and moles that form a monophyletic group apart from marsupial wolves.

Figure 6.4 abstracts away from the six taxa that were used as examples in Figure 6.3. Note that the horizontal dimension no longer represents overall dissimilarity. Consider the three taxa (*A*, *B*, and *C*) depicted in (a). Their most recent common ancestor is depicted at the root of the two trees. The character

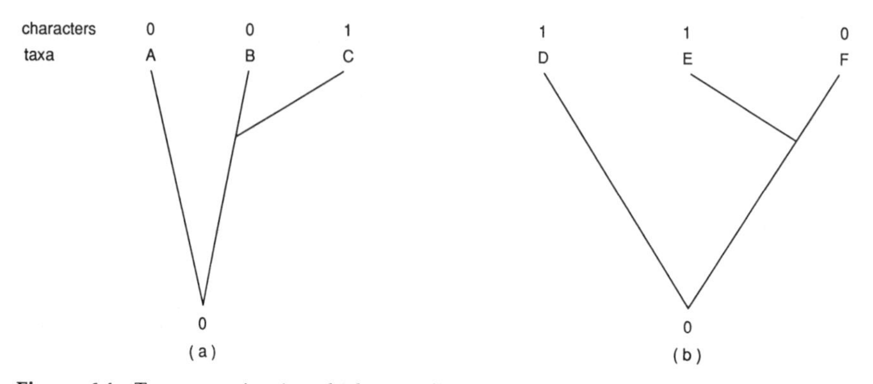

Figure 6.4 Two scenarios in which overall similarity can fail to reflect genealogical relatedness. In (a), *A* and *B* retain ancestral characters (0), whereas *C* evolves a set of evolutionary novelties (1). In (b), *D* and *E* independently evolve the same set of evolutionary novelties (1), whereas *F* retains the ancestral condition (0).

state of this ancestor is denoted by 0. In pattern (a), the descendants *A* and *B* retain the ancestral condition, while *C* has evolved a novel characteristic (denoted by 1). The monophyletic grouping is A(BC), although this is not reflected by the overall similarity of the three taxa. In pattern (b), *D* and *E* independently evolved the same novel condition, while *F* retained the ancestral state. Here again, overall similarity is a misleading guide to propinquity of descent.

Another way to characterize the difference between patterns (a) and (b) is worth mentioning. When two species share a characteristic because they inherited it unmodified from a common ancestor, the similarity is said to be a *homology*. Alternatively, when two species share the characteristic because it evolved independently in the lineage leading to one and in the lineage leading to the other, the similarity is said to be a *homoplasy*.

This distinction applies to Figure 6.4 as follows. The resemblance between *A* and *B* in (a) could very well be a *homology*. Figure 6.4(a) does not say that this must be so; after all, it is possible that the trait flip-flopped from 0 to 1 and back again on the lineages leading from the ancestor to *A* and to *B*. But pattern (a) is quite consistent with the hypothesis that the similarity of *A* and *B* is a homology. The same cannot be said of pattern (b) in Figure 6.4. If *D* and *E* have state 1 and their most recent common ancestor was in state 0, then the novel trait must have originated twice. The similarity of *D* and *E* cannot be a homology; it must be a homoplasy.

Pheneticists do not care about this distinction. Based just on overall similarity, they classify *A* and *B* together apart from *C* for the same reason that they classify *D* and *E* together apart from *F*. They go by overall similarity; whether that similarity reflects homologies or homoplasies is of no interest to them.

This leads me to the third taxonomic philosophy, which, conceptually, is a kind of compromise between cladistics and phenetics. This is the approach called *evolutionary taxonomy*. Evolutionary taxonomists permit similarity to override genealogy *sometimes but not always*. They put lizards and crocodiles together apart from birds. But they refuse to put marsupial wolves and placental wolves together apart from moles. The adaptations shared by lizards and crocodiles are homologous, which is why evolutionary taxonomists place them together apart from the birds (which evolved into a new adaptive niche). In contrast, the adaptive similarities of marsupial and placental wolves are not homologies; in this case, one classifies by genealogy, not by similarity.

So cladists, pheneticists, and evolutionary taxonomists all believe that classification should be hierarchical. However, their reasons for requiring this are quite different. Cladists advocate hierarchy because they want classification to reflect precisely the evolutionary branching process. Pheneticists favor hierarchical classifications because that is how they wish to describe nested similarity relationships. And evolutionary taxonomists want a hierarchical classification because this is the structure that best represents their preferred mixture of branching structure and adaptive similarity.

This difference in philosophy makes a real difference in the taxa that each school recognizes as biologically real. As already noted, cladists reject *Reptilia,* though pheneticists and evolutionary taxonomists consider it genuine. Cladists also reject *Invertebrata, Acrania, Pisces,* and a number of other traditional taxa. It is a mistake to think that systematists through the ages have always identified the same basic pattern of natural order: The cladistic focus on branching pattern involves a genuine revolution.

My presentation of these three taxonomic philosophies left evolutionary taxonomy until last since it is a conceptual compromise between cladistics and phenetics. However, this logical ordering is untrue to the historical order in which the three positions developed. Evolutionary taxonomy is the ancestral condition—the oldest of the three viewpoints, an outgrowth of the Modern Synthesis. It is typified by the work of Mayr (1942) and Simpson (1961). Pheneticists rebelled against this reigning orthodoxy in the 1960s. One of pheneticism's central documents is the book by Sneath and Sokal (1973). Pheneticists argued that the evolutionary scenarios about adaptation on which evolutionary taxonomists based their classifications were not well supported by objective evidence. Pheneticists sought to make their methods explicit, so that computers could generate classifications from the data of similarity and difference. By forcing one's methodology out in the open, it would be impossible to appeal to "intuition," which pheneticists felt was a dodge for defending an ill-conceived orthodoxy. They wanted to eliminate evolutionary theory from the enterprise of systematics because they believed that the introduction of theory into systematics would rob the science of its objectivity (Hull 1970).

Cladism arose as a revolt against both pheneticism and the older school of evolutionary taxonomy. This rebellion came into its own (at least in the United States) in the 1970s, once the 1966 English translation of Hennig's 1950 book became widely known. Cladists shared with pheneticists a dissatisfaction with what they thought was the subjectivity of evolutionary taxonomy. However, their goal was not to get rid of evolutionary theory *in toto* but to focus on the branching pattern of evolution as the one true foundation on which systematics could be constructed.

6.4 Internal Coherence

In many cases of theory choice in science, the competing theories aim to characterize (roughly) the same class of phenomena. One then evaluates the theories by seeing which of them does the best job. Conversely, when two theories address different problems, there is no reason to choose between them. It would be pointless, after all, to ask whether Keynesian economics is preferable to plate tectonics. The theories have different goals, so why think that they compete with each other?

One peculiarity of the dispute about classification described in the previous section is that the different systematic philosophies have different goals. One might paper over these differences by saying that they all aim to formulate "the natural classification" or "to systematize life's diversity." But, in fact, cladistics focuses exclusively on branching pattern, phenetics on overall similarity, and evolutionary taxonomy on a mixture of branching pattern and adaptive similarity. How is the dispute among these schools to be resolved, if they, in fact, have different fundamental goals?

One possible response is to regard the choice of systematic philosophy as a matter of convention. Many scientists outside of systematics seem to have this attitude. They view classifications as more or less arbitrary ways to catalog life's diversity. Just as it is convenience, not truth, that dictates how the books in a library should be arranged, so there can be no factual basis for saying that one classification is true and another false.

However, before we can adopt this conventionalist view, there is an issue we must investigate: We must see if each approach is *internally coherent*. Only if this is true will conventionalism be a viable position.

Many defenses of pheneticism adopt a very narrow empiricist attitude to science. It is claimed that a truly "objective" classification must be based on "pure" or "direct" observation, untainted by the influences of theory—especially of evolutionary theory. Hull (1970) persuasively argues that theory neutrality of this sort is not only undesirable but also is impossible to achieve. When a biologist claims that two organisms exhibit "the same" characteristic, this assertion goes beyond the raw data of experience—every two organisms exhibit myriad similarities and myriad differences. Classification inevitably involves selectively labeling some of those similarities and calling them "characters." Observational data are influenced by theoretical considerations about what counts as salient.

Pheneticists also have argued that it is impossible to ascertain genealogical relationships and that classifications therefore cannot be based on reliable evolutionary information. If evolution really were unknowable, that *would* undermine cladistics and evolutionary systematics. Indeed, we will see in the next two sections that there is genuine controversy in evolutionary theory about how phylogenetic relationships should be inferred. Nonetheless, it is an exaggeration to claim that genealogy is unknowable. Granted, it can never be known *with certainty.* And granted again, there are taxa whose phylogenetic relationships are controversial. But there are many taxa about whose branching pattern virtually all biologists agree. Pheneticism cannot win by default—by claiming that it is the only game in town.

A third argument pheneticists have produced for their philosophy contends that classifying by overall similarity yields an "all-purpose classification." They concede that cladistic and evolutionary taxonomies may be useful in specifically evolutionary contexts, but they insist that classification can and should aspire to a more general utility.

It clearly makes sense to assess the merits of a classification once a specific purpose has been described with sufficient precision. But to talk of an "all-purpose classification" is to say something about the merits of a classification relative to *all possible purposes.* It is entirely unclear how one should understand this variegated set of aims and how the overall merits of a classification are to be gleaned from the merits and demerits it has with respect to each possible purpose. A saw is better for cutting wood than a hammer; a hammer is better for pounding nails than a saw. But how is one to compare a saw and a hammer as "all-purpose tools"? Pending further clarification, the idea of an "all-purpose classification" should be dismissed as unintelligible.

I so far have identified three bad arguments for pheneticism. These arguments to one side, what can be said directly about that philosophy? Can these bad arguments be replaced by solid defenses of pheneticism, or is there something basically wrong with the whole approach?

I suggest that there is a fundamental difference between cladistics and phenetics. Cladistic classification aims to record *the set of monophyletic groups.* It may be difficult or easy to figure out what those monophyletic groups are. But, out there in nature, there is a unique branching structure into which taxa fit. The goal of cladistic classification is quite intelligible. Whether that goal is attainable, of course, is another matter.

The question I want to pose about phenetics is whether, in nature, there is a determinate fact concerning how overall similar different pairs of taxa really are. The goal of phenetic classification is to record overall similarity, but is there really such a thing to begin with? I am inclined to doubt that this is an objective feature of reality.

If I enumerated a set of properties to be considered and stipulated a weighting scheme to be followed, then we could calculate how similar different pairs of objects are. If I tell you to consider just sex and hair color and to accord these equal weight, then you can say that brown-haired Alice and Betty are more similar to each other than either is to redheaded Carl. What I doubt is that there is such a thing as the overall similarity uniting pairs of individuals. The idea of *overall* similarity purports to take account of *all* the properties that each individual possesses. I have no idea what this totality is supposed to be. Overall similarity also requires some scheme for weighting the characters in this totality; equal weighting is one scheme, but why favor equal weighting over some other procedure?

Consider, for example, human beings, chimps, and snakes. A morphologist can point to numerous traits that humans and chimps have in common but that snakes do not possess. However, for each of these characteristics, it is possible to describe others that entail a quite different similarity grouping. True, humans and chimps are warm blooded, while snakes are not. On the other hand, humans speak a language, but chimps and snakes do not, and chimps are covered with fur, whereas humans and snakes are not. If every description we formulate is taken to pick out a characteristic, then it is hard to resist the conclusion that there are as many characteristics that unite one pair of these as unite any other.

So the first problem for pheneticism is the problem of saying what counts as a character. But there is a second difficulty, one that has been elucidated by pheneticists themselves. Their admirable passion for precision has led them to describe a large number of similarity measures. By their own admission, there is no such thing as *the* overall similarity that characterizes the data one has on different pairs of objects. A set of species can be organized into different taxonomies, depending on which measure of similarity one applies to the data at hand (Johnson 1970; Ridley 1986).

I turn now to the defense of evolutionary taxonomy. Mayr (1981) argues that evolutionary classifications are preferable to cladistic classifications because the former are *more informative* than the latter. His ground for saying this is that evolutionary taxonomies reflect both branching pattern *and* character divergence, whereas cladistic taxonomies reflect branching pattern alone. The problem with this argument is that it conflates *input* with *output;* it confuses the information taken into account when one constructs a classification with what the classification entails, once it is formulated (Eldredge and Cracraft 1980; Ridley 1986). More informative classifications *say more.* However, it is entirely unclear why grouping crocodiles with lizards but apart from birds provides more information than putting crocs and birds together apart from lizards.

What does it mean to compare statements for their "informativeness"? Informativeness is relative to the question asked. For example, "Sam is tall" says more about Sam's height than "Sam has dark hair" does, but the latter tells you more about his hair color than the former does. Neither sentence is more informative in any absolute sense. My suspicion is that cladistic classifications are most informative if it is monophyly that you want to know about. However, phenetic classifications are most informative if overall similarity is what you hanker after. And if it is the mixed information codified by evolutionary taxonomies that you seek, then evolutionary taxonomies will deliver that best of all.

In the end, the fatal flaw in evolutionary taxonomy is that it has never been able to formulate a nonarbitrary criterion for when homology matters more than propinquity of descent. Crocodiles and lizards share a number of homologies that birds do not possess, but it also is true that crocs and birds share homologies that distinguish them from lizards. Why are the former homologies so much more important than the latter that we should decline to classify by propinquity of descent? If lizards and crocs had been less similar and crocs and birds had been more so, evolutionary taxonomists would have promulgated a different proposal. Where on this sliding scale does similarity give way to genealogical relatedness? This fundamental problem for evolutionary taxonomy seems insoluble.

6.5 Phylogenetic Inference Based on Overall Similarity

In the previous two sections, I described the three systematic schools as giving different answers to the following question: If you knew the phylogenetic tree for a set of species, how would you use that information to construct a classification? Cladists would take that information to suffice for classification; evolutionary taxonomists would regard the information as relevant, though incomplete; and pheneticists would say that the information is of the wrong sort.

So the three schools disagree about the relationship of phylogenetic branching pattern and classification. However, there is a prior question that has been of major importance in systematics: How is one to infer what the phylogenetic tree is for a set of taxa? Here, there are basically two positions, with several variants within each. There is the phenetic approach and the cladistic approach to phylogenetic inference. I'll discuss the former now and leave cladistics for the next section.

Are human beings and chimps more closely related to each other than either is to lions? That is, do human beings and chimps have an ancestor that is not an ancestor of lions? If so, humans and chimps belong to a monophyletic group that does not include lions. How is one to decide whether the evolutionary branching process proceeded in this way? The phenetic approach to this genealogical problem is to collect data on the three species and to calculate which two of them are most similar. Humans and chimps would be judged to form a group apart from lions if they turn out to be more similar to each other than either is to lions.

Although the use of overall similarity to infer phylogenetic relationships may have some *prima facie* intuitive appeal, there is a problem that advocates of this procedure must address. Figures 6.3 and 6.4 depict two evolutionary situations in which overall similarity fails to reflect genealogy. When the taxa under study evolve by a branching process of pattern (a) or pattern (b), phylogenetic inference based on overall similarity will lead to a false picture of genealogy. How do we know that the pattern of similarity in the taxa we are studying is not misleading in one of these two ways? The question is not whether we are entitled to be absolutely certain about genealogy. Let us grant that absolute certainty is impossible in science. The question is why we should think that overall similarity is a reasonable guide to phylogenetic relationships. Sneath and Sokal (1973, p. 321) offer the following answer:

> Those who would largely use phenetic similarity as evidence of recency of cladistic ancestry must assume at least some uniformity of evolutionary rates in the several clades. ... We stress here uniform rather than constant rates of evolution. As long as rates of evolution change equally in parallel lines, it is unimportant whether these rates are constant through an evolutionary epoch. We may use the analogy of multiple clusters of fireworks, smaller clusters bursting from

inside larger clusters, a familiar sight to most readers. While the small rays of the rocket have not "evolved" at all until the small rocket explodes, their rates of divergence from the center of their rocket are identical but not constant, since they were zero during the period of the early ascent of the rocket.

Figure 6.5 illustrates what Sneath and Sokal have in mind. Suppose this tree depicts the true phylogenetic relationship that obtains among *A, B,* and *C*. I have divided the evolution of the three taxa into two temporal periods. The first goes from the tree's root to the point at which the lines leading to *A* and *B* diverge; the second extends from that branching point to the present.

Let us consider a single character that evolves in this tree; suppose, for simplicity, that it comes in just two states (0 and 1). For each of the branches represented in this figure, we can define two *transition probabilities:*

$$e_i = P(\text{the } i\text{th branch ends in state 1 / the } i\text{th branch begins in state 0})$$
$$r_i = P(\text{the } i\text{th branch ends in state 0 / the } i\text{th branch begins in state 1})$$

These letters are chosen as mnemonics for *evolution* and *reversal.*

What relationships must obtain among the various e_i's and r_i's for it to be probable that the method of overall similarity will recover the true genealogy? Sneath and Sokal's answer is that branches 1, 2, and 3 must be characterized by approximately the same *e*'s and *r*'s and that branches 4 and 5 must be characterized by approximately the same transition probabilities as well. This is what they mean by *uniform* rates: The rules of character evolution are the same for branches *within the same temporal period.* Uniformity does not demand *constancy* (i.e., that branches in *different periods* obey the same rules of change).

Let us formulate Sneath and Sokal's argument more carefully. Since there are only three taxa in the example considered, there are eight possible distributions that a given character might have:

	A	*B*	*C*
1.	1	1	1
2.	1	1	0
3.	1	0	1
4.	0	1	1
5.	0	0	1
6.	0	1	0
7.	1	0	0
8.	0	0	0

Patterns 1 and 8 are useless as far as the method of overall similarity is concerned since they do not separate two of the taxa from the third. Characters that conform to patterns 2 and 5 would lead the method of overall similarity to infer correctly that the true genealogy is (*AB*)*C*. Patterns 3 and 6 point to (*AC*)*B*, and patterns 4 and 7 point to *A*(*BC*). Notice that the method of overall similarity will correctly infer that (AB)C is the true phylogeny from a set of

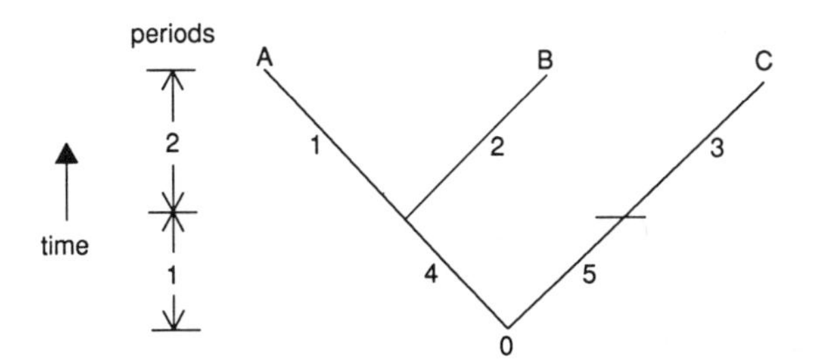

Figure 6.5 Character evolution obeys the assumption of *uniform rates* if contemporaneous lineages obey the same rules of change.

data only if the data set contains more characters of patterns 2 and 5 than characters that conform to patterns 3 and 6.

We would like to know when it is rational to use the method of overall similarity to infer genealogical relationships. For ease of exposition, I'll denote pattern 2 by "110," pattern 3 by "101," etc. Assuming that (*AB*)*C* is the true phylogeny, we can state Sneath and Sokal's proposal in two parts:

(1) It is rational to use the method of overall similarity if and only if *P*(110 or 001 / (*AB*)*C*) > *P*(101 or 010 / (*AB*)*C*), *P*(011 or 100 / (*AB*)*C*).

(2) *P*(110 or 001 / (*AB*)*C*) > *P*(101 or 010 / (*AB*)*C*), *P*(011 or 100 / (*AB*)*C*) only if rates of evolution are approximately uniform.

Proposition (1) characterizes the *condition of adequacy* that I take Sneath and Sokal to be imposing. Proposition (2) states that approximate uniformity of rates is needed for that condition of adequacy to be satisfied. I will not consider proposition (2) any further but will leave it to the reader to figure out if it is true.

Sneath and Sokal do not explicitly formulate a condition of adequacy. However, it is not implausible to think that (1) is implicit in their argument. Suppose we look at a few characters drawn from the three taxa we are considering. If the condition described in (1) is true for each of the characters we examine, then *A* and *B* will probably be more similar to each other than either is to *C*. If we examine larger and larger data sets, then the probability approaches unity that *A* and *B* will be the most similar pair (by the Law of Large Numbers; see Section 3.2).

Proposition (1) says that it is rational to use the method of overall similarity precisely when that method *would not be misleading in the long run*. We are to evaluate the method of inference in terms of this asymptotic property. The argument judges the method to be reasonable if it would converge on the truth were the data set made large without limit.

It is an interesting philosophical question whether convergence (which statisticians call *consistency*) is *necessary* for a rule of inference to be reasonable (Sober 1988). I won't consider that issue here but instead will focus on whether convergence *suffices* to justify a rule of inference. That is, let us suppose that overall similarity would converge on the true phylogeny if supplied with infinite data. Is that enough to justify using the method on the finite data sets that systematists actually confront?

I think the answer is no. There are other methods for inferring phylogenetic relationships besides overall similarity. One of them, which I'll discuss in the next section, is cladistic parsimony. These two methods often disagree about how to interpret the *finite* data sets that systematists consider in real life. The fact that overall similarity will converge on the truth in the infinite long run leaves unanswered the question of whether cladistic parsimony may do the very same thing. If the data set were infinitely large, perhaps the two methods would agree. However, this does not help us decide which method to use when the data set before us is finite and the methods *dis*agree.

This scenario is not hypothetical. For three taxa, overall similarity converges on the truth with infinite data if the assumption of uniform rates is correct; however, cladistic parsimony recovers the true phylogeny in this circumstance as well (Sober 1988).

So the fact that a method would converge on the truth in the infinite long run is not enough to justify using that method on finite data sets. This point is a familiar one in statistical reasoning. Suppose we wish to infer the mean height of the individuals in a population in which height is normally distributed. We sample with replacement. In the infinite long run, the *mean* of the sample must coincide with the mean of the population. However, it also is true in the infinite long run that the *mode* of the sample (the most frequently exemplified height) must coincide with the mean of the population. So, with *infinite* data, using the mean and using the mode will lead to the same (true) answer. The problem is that, with *finite* data, using the mean can lead to an estimate that differs from the one obtained by using the mode. The fact that a method converges in the long run does not suffice to justify that method.

The second problem with the appeal to uniform rates is one that Sneath and Sokal (1973, p. 321) mention: "Clearly all the evidence at hand indicates evolutionary rates in different clades are not uniform. Different lines do evolve at different rates." If the rationality of using overall similarity really *depends* on approximate uniformity of rates and if the uniform rate assumption is false, then it must be *ir*rational to use overall similarity.

Even if the uniform rates assumption is not realistic in general, there still may be special cases in which it makes sense. For example, consider molecular characters that evolve by random drift. The rate of substitution under drift is determined by the mutation rate (not by population size). If the different taxa in a clade have approximately the same mutation rates, uniformity of rates may be a reasonable assumption. However, it needs to be emphasized that, even here, uniformity is not enough. Even if overall similarity satisfies the requirement of convergence, the question remains of whether *other methods* (like cladistic parsimony) do so as well.

Before I consider cladistic parsimony, a couple of general conclusions can be drawn. The first is that every argument about the rationality (or irrationality) of a method of phylogenetic inference must include a condition of adequacy. If the question is whether it is rational to use a method, we must say what we mean by "rational." Earlier in this section, I reconstructed Sneath and Sokal's answer to this question as proposition (1).

The second lesson is that once a condition of adequacy has been specified, whether a method of inference satisfies this condition will depend on facts about the evolutionary process. Overall similarity may strike one as a very intuitive guide to propinquity of descent. But this appeal to intuition is not enough; once the argument is stated clearly, the need for evolutionary assumptions of some sort becomes palpable. Proposition (2) enters the story at this point.

To understand the issues here, it is important not to confuse the problem of phylogenetic inference with the problem of classification (Felsenstein 1984). Conceivably, someone might use phenetic methods for reconstructing phylogenetic trees but then use nonphenetic (e.g., cladistic) methods for constructing a classification. The opposite mixture is conceivable as well—use cladistic parsimony to obtain a tree but then use noncladistic methods in classification. In fact, evolutionary taxonomists often advocate the latter sort of mixed approach (Mayr 1981). Systematic philosophies have at least two parts. The different parts must be assessed separately.

6.6 Parsimony and Phylogenetic Inference

The idea that the principle of parsimony is relevant to the project of phylogenetic inference has been implicit in evolutionary theorizing for a long time. Arguably, appeal to parsimony lies behind every hypothesis of homology; it is more parsimonious to explain (some) similarities between species by postulating a common ancestor from which the trait was inherited than to postulate two independent originations (Nelson and Platnick 1981).

Consider the problem depicted in Figure 6.6. Robins and sparrows both have wings. How might we explain this similarity? One possibility is to postulate a winged common ancestor from which the trait was inherited. This hypothesis says that the similarity is a *homology*. An alternative is to say that the trait evolved twice—once in the lineage leading to robins and once in the lineage leading to sparrows. This hypothesis treats the similarity as a *homoplasy*. The first hypothesis appears to be the more parsimonious explanation since it postulates fewer origination events.

This parsimony argument seems to have a likelihood rationale. To see why, consider a nonbiological example. Two students hand in essays in a philosophy course that are word-for-word identical. Two hypotheses might be considered to explain the similarity. The first is a *common cause explanation*; it says that the students copied from a paper they found in a fraternity house file. The second is a *separate cause explanation*; it says that the students worked independently and just happened to produce the same essay.

Which of these explanations is more plausible? The Likelihood Principle (Section 2.2) has an answer. What is the probability of the essays matching if the common cause (*CC*) hypothesis were true? What is the probability of their matching, if the separate cause (*SC*) explanation were true? Whatever the ex-

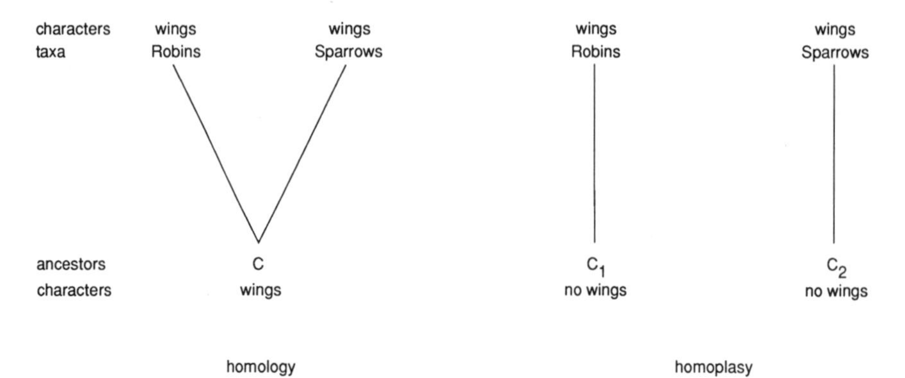

Figure 6.6 Two possible explanations of why Robins and Sparrows both have wings. According to the hypothesis of *homology,* they inherited the trait from a common ancestor. According to the hypothesis of *homoplasy,* the trait evolved independently in the two lineages.

act values happen to be, there seems to be a drastic *difference* in these probabilities: *P*(the essays match / *CC*) >> *P*(the essays match / *SC*). The Likelihood Principle tells us that the evidence strongly favors the common cause hypothesis.

The common cause hypothesis postulates a single cause. The separate cause hypothesis postulates two quite separate causal processes. The common cause hypothesis seems to be more parsimonious because it postulates fewer causal processes. In this example, the principle of parsimony seems to have a likelihood rationale.

Even though considerations of parsimony have long been implicit in evolutionary hypotheses of homology, it was not until the 1960s that a principle of parsimony aimed specifically at the problem of phylogenetic inference was stated explicitly. The principle received two rather independent formulations. First, there was the work of Edwards and Cavalli-Sforza (1963, 1964), students of R. A. Fisher, who approached phylogenetic inference as a statistical problem. They recommended parsimony as a *prima facie* plausible inference principle, whose ultimate justification they regarded as not immediately obvious. Second, there was Willi Hennig's (1966) book and its subsequent elaboration by a generation of cladists. Hennig did not use the term "parsimony," but the basic ideas that he developed were widely interpreted in that idiom.

Whereas Edwards and Cavalli-Sforza were rather circumspect about whether and why parsimony is a reasonable rule of inference, Hennig and his followers formulated arguments to show how and why a principle of parsimony should be used in phylogenetic inference. I'll now describe Hennig's (1965, 1966) basic line of reasoning; elaborations of it may be found in Eldredge and Cracraft (1980) and in Wiley (1981).

A properly formulated problem of phylogenetic inference involves at least three taxa. For example, one doesn't ask whether robins and sparrows are re-

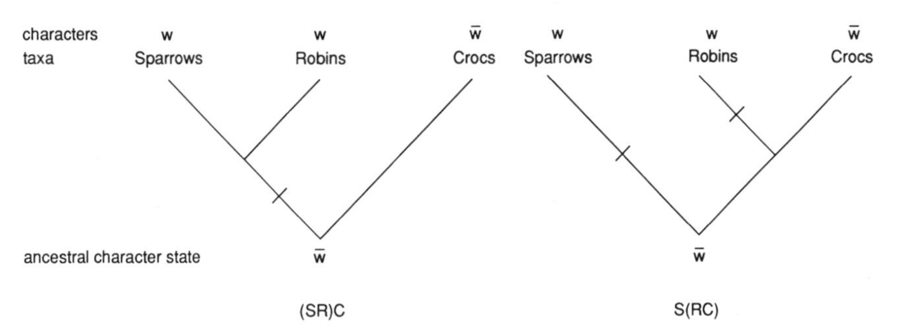

Figure 6.7 Sparrows and Robins have wings, whereas Crocs do not. If winglessness is the ancestral condition, then the genealogical hypothesis (SR)C requires a single change in character state to explain the observations, whereas S(RC) requires two.

lated to each other. Rather, the problem is to say whether robins (*R*) and sparrows (*S*) are more closely related to each other than either is to, say, crocodiles (*C*). The three competing hypotheses are (*SR*)*C*, *S*(*RC*), and (*SC*)*R*. Each says that the two bracketed taxa belong to a monophyletic group that does not include the third. Two of these hypotheses are depicted in Figure 6.7. Each assumes that the three taxa have a common ancestor if we go back far enough in time.

Let's consider how the distribution of a single character is relevant to deciding which hypothesis is more plausible. Sparrows and robins have wings; crocodiles do not. What is the most parsimonious explanation of this pattern? Suppose, for the sake of argument, that the most recent ancestor common to all three taxa lacked wings. That is, suppose we know how to *polarize* the character—wingless is the ancestral (*plesiomorphic*) and winged is the derived (*apomorphic*) character state. This assumption is depicted in Figure 6.7. We now can ask what pattern of character evolution must have taken place in the interior of a tree, if that tree is to produce the character distribution we observe at the tips.

For the (*SR*)*C* tree to produce the characters at the tips, only a single change in character state need take place in the tree's interior. A single change from wingless to winged, on the branch in (*SR*)*C* through which a slash has been drawn, suffices to generate the pattern we observe at the tips. However, for *S*(*RC*) to generate the observations, at least two changes (each depicted by a slash) must occur. Since (*SR*)*C* requires only one change while *S*(*RC*) requires at least two, we conclude that (*SR*)*C* is the more parsimonious explanation of the data. The principle of parsimony concludes that the data favor (*SR*)*C* over *S*(*RC*).

I hope the relationship between Figures 6.6 and 6.7 is clear. (*SR*)*C* is the more parsimonious hypothesis, and it is consistent with the idea that the wing shared by sparrows and robins is a homology. *S*(*RC*) is the less parsimonious hypothesis. It is *in*consistent with the hypothesis of homology; if *S*(*RC*) is true, then the resemblance *must* be a homoplasy.

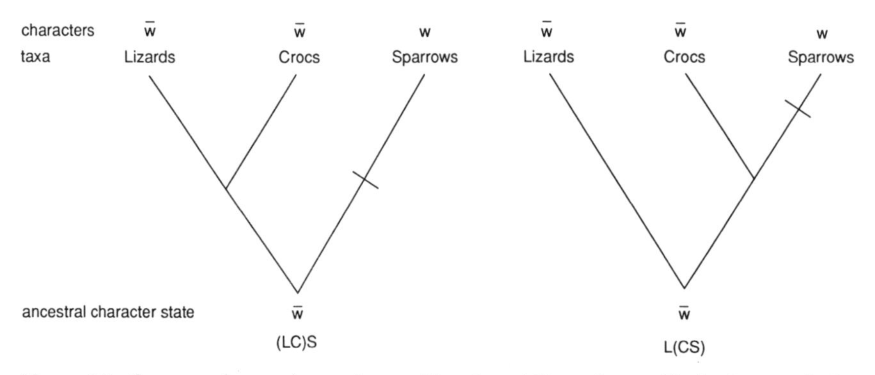

Figure 6.8 Sparrows have wings, whereas Lizards and Crocs do not. If winglessness is the ancestral condition, then the genealogical hypotheses (LC)S and L(CS) each require a single change in character state to explain the observations.

So far, it may seem that considerations of parsimony agree with considerations of overall similarity. (*SR*)*C* is the more parsimonious hypothesis relative to the one character just described; it also is the hypothesis favored by considerations of overall similarity, again relative to this one character. We now need to see how parsimony and overall similarity can conflict.

Consider Figure 6.8. The problem is to infer the pairwise grouping for lizards, crocodiles, and sparrows, given the single character distribution depicted. Again, one needs an assumption about character polarity to calculate which hypothesis is more parsimonious. Assuming that winglessness is the ancestral condition for these three taxa, we now can ask how many changes must occur between root and tips for each tree to produce the observed character distribution. The answer is that (*LC*)*S* can generate the observations with only a single change in the tree's interior, *and the same is true of L(CS)*. The trees are equally parsimonious, so the principle of parsimony concludes that the data do not discriminate between the two hypotheses.

In Figure 6.7, the similarity uniting sparrows and robins is a *derived similarity* (a *synapomorphy*). In Figure 6.8, the similarity uniting lizards and crocodiles is an *ancestral similarity* (a *symplesiomorphy*). This pair of examples illustrates how these two kinds of similarity differ in significance when the principle of parsimony is used to infer phylogenetic relationships. Derived similarities count as evidence of relatedness because they make a difference in how parsimonious the competing hypotheses are. In contrast, ancestral similarities often fail to count as evidence of relatedness because they often fail to make a difference in how parsimonious the competing hypotheses are. Cladistic parsimony is a different method of phylogenetic inference from overall similarity because cladistic parsimony refuses to count *all* similarities as evidence.

Cladists, unlike pheneticists, think that ancestral and derived similarities differ in their evidential significance. But how is one to tell whether a similarity is of one sort or the other? The standard cladistic procedure for determin-

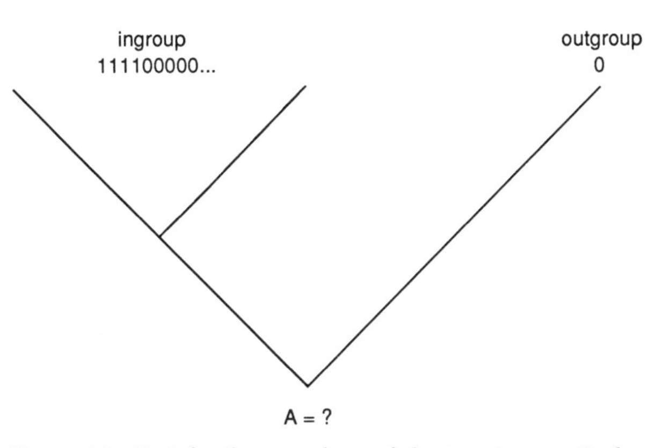

Figure 6.9 To infer the genealogy of the taxa in a particular group (the "ingroup") by using the method of cladistic parsimony, one must *polarize* the characters—i.e., infer for each character which of its states is ancestral and which is derived. The *method of outgroup comparison* says, roughly, that one should find a taxon outside the group of interest and use the character state of this "outgroup" as the best estimate of the ancestral character state.

ing character polarity is *the method of outgroup comparison*. Consider the taxa depicted in Figure 6.9. We want to infer the phylogenetic relationships that obtain among the taxa in the *ingroup*. The character we are considering comes in two states—0 and 1. How should this character be polarized? I have assumed arbitrarily that 0 = ancestral and 1 = derived, but how are we to find out which is which, in fact? The method of outgroup comparison bids us find a taxon that is outside the group of interest. The method says that the best estimate of the ancestral condition is the character state of this outgroup. Since the outgroup is in state 0, we should infer that 0 is the ancestral condition in the ingroup. (This is a simplified rendition of the procedure; see Maddison, Donoghue, and Maddison 1984 for further discussion.)

There is an intuitive connection between the outgroup comparison method and parsimony. If we say that the ancestor *A* is in state 0, no change is required on the branch leading from *A* to the outgroup. Alternatively, if we say that *A* is in state 1, a change must occur on that branch. But regardless of how we polarize the character, there must be at least one change in character state within the ingroup. So the hypothesis that *A* is in state 0 is more parsimonious than the hypothesis that *A* is in state 1.

Thus far, I have described some of the basics about cladistic parsimony by concentrating on what that method says about single character distributions. However, phylogenetic inference typically is based on data sets containing a great many characters. If the task is to find the best tree for the three taxa *A*, *B*, and *C*, some of the characters may be synapomorphies with the 110 pattern; these will favor (*AB*)*C*. Other, conflicting characters may exhibit the 011 pat-

Box 6.2 "Defining" Monophyletic Groups

Essentialism says that a taxon can be defined by describing traits that all and only its members possess (Section 6.1). Isn't it evidence for essentialism that higher taxa are defined in terms of various characters? Organisms in *Vertebrata* have backbones; organisms in *Aves* have feathers. Don't these statements describe the essences of the taxa described?

No. If a taxon is monophyletic, then it includes *all* the descendants of some ancestral species. A descendant of a species in *Vertebrata* must be a member of *Vertebrata*, whether or not it has a backbone. And a featherless descendant of a bird species is still a bird, if "bird" names a monophyletic group. By demanding that taxa be monophyletic, cladistics has anti-essentialist consequences.

Characters do not "define" biological taxa (if a definition must provide necessary and sufficient conditions). A trilateral is defined as a closed plane figure with three straight sides. If something fails to have any of the defining characteristics, it fails to be a trilateral. Characters *are evidence for* the existence of a monophyletic group. The fact that sparrows and robins have wings, but snakes do not, is evidence that sparrows and robins belong to a monophyletic group that does not include snakes. In addition, characters *mark the origination* of new monophyletic groups. When birds first evolved, feathers were a novelty.

tern; these are most parsimoniously explained by the *A*(*BC*) hypothesis. The overall most parsimonious hypothesis is the one that minimizes the number of homoplasies needed to explain *all* the data. Cladists frequently, though not always, accord equal "weight" to each character. Each required homoplasy is taken to diminish a hypothesis's plausibility to the same degree.

Since its original formulation, the justification for using cladistic parsimony to infer phylogenetic relationships has been controversial. Hennig (1966, pp. 121–122), in part quoting himself, espoused the following "auxiliary principle":

> that the presence of apomorphous characters in different species "is always reason for suspecting kinship (i.e., that the species belong to a monophyletic group), and that their origin by convergence should not be assumed a priori." . . . That was based on the conviction that "phylogenetic systematics would lose all the ground on which it stands," if the presence of apomorphous characters in different species were considered first of all as convergences (or parallelisms), with proof to the contrary required in each case. Rather the burden of proof must be placed on the contention that "in individual cases the possession of common apomorphous characters may be based only on convergence (or parallelism)."

Hennig's remarks raise two questions. Even if phylogenetic inference would be impossible without the assumption that shared derived characters are evidence of relatedness, why does that show that the assumption is a reasonable one? Furthermore, even if we grant that shared *derived* characters are evidence of relatedness, why should we also think that shared *ancestral* characters are evidentially meaningless?

More recently, several cladists have tried to justify parsimony by using Popper's ideas about falsifiability (Gaffney 1979; Eldredge and Cracraft 1980; Nelson and Platnick 1981; Wiley 1981; Farris 1983). They have argued that a genealogical hypothesis is falsified each time it is forced to say that a character evolved more than once. They conclude that the least falsified hypothesis is the one that requires the fewest homoplasies.

To evaluate this idea, we need to consider the concept of falsification. On one reading of that concept (Popper's), "*O* falsifies *H*" means that if *O* is true, *H* must be false. An observation falsifies a hypothesis, in this sense, if one can *deduce* that the hypothesis is false from the truth of the observation statement (Section 2.7). It is quite clear that character distributions are not related to genealogical hypotheses in this way. Thus, the hypothesis (*AB*)*C* is logically consistent with all possible observations. The 011 character distribution does *not* falsify (*AB*)*C* in Popper's sense of falsification.

Another reason for rejecting this deductivist reading of what falsification means is that even the most parsimonious tree constructed from a given data set usually requires some homoplasies. It requires fewer than its competitors, but that does not mean that it requires none at all. If a tree literally were falsified by the homoplasies it requires, then even the most parsimonious tree would have to be false.

Clearly, a *non*deductive reading of falsification is needed. What is really being asserted is that a hypothesis is *disconfirmed* (i.e., its plausibility is reduced) each time it requires a homoplasy. The most parsimonious hypothesis usually requires *some* homoplasies. It has some strikes against it, but it has fewer strikes against it than the competing hypotheses have.

We now must return to the questions raised in connection with Hennig: Why is the plausibility of a hypothesis reduced each time it requires a homoplasy? And why is this the only property of the data that is relevant to evaluating competing hypotheses? After all, a pheneticist could agree that a 110 character distribution favors (*AB*)*C* over *A*(*BC*). The question is why a 001 character distribution does not have the same evidential meaning.

Popper argued that it is intrinsic to the scientific method to regard parsimony as a virtue of scientific hypotheses. For him, the principle of parsimony is part of the logic of science; it is a quite general principle of reasoning whose justification is independent of the specific subject matters to which it is applied. Cladists have followed Popper by arguing that it is part of the scientific method that more parsimonious *genealogical* hypotheses should be preferred over their less parsimonious competitors. A central idea in cladistic writing about parsimony is that *using parsimony to infer genealogical patterns does not require substantive assumptions about the evolutionary process* (see, for example, Farris 1983).

The opposite point of view has been defended by Felsenstein. He has argued that when parsimony is evaluated in the light of various statistical criteria, the adequacy of the method depends on specific assumptions about the evolutionary process. For example, Felsenstein (1978) shows that parsimony will converge on the truth in the infinite long run in many but not all evolutionary situations. And Felsenstein (1983) argues that the most parsimonious hypothesis will be the hypothesis of maximum likelihood in some evolutionary circumstances but not in others.

Cladists sometimes dismiss Felsenstein's arguments by claiming that the examples he discusses are not realistic. This reaction would be legitimate if his point was to show that parsimony should not be used to make phylogenetic inferences. On the other hand, if the point is to demonstrate that the use of parsimony requires evolutionary assumptions, Felsenstein's arguments cannot be faulted on this ground.

What would it mean to evaluate the credentials of cladistic parsimony from the point of view of likelihood? As in the discussion of Sneath and Sokal's proposal in the previous section, two claims would have to be defended:

(1) The most parsimonious explanation of the data is the most plausible explanation if and only if that explanation is the one with the highest likelihood.

(2) The most parsimonious explanation of the data is the hypothesis of maximum likelihood only if evolution has property X.

Proposition (1) says that likelihood has been adopted as the criterion of adequacy; once X is specified, proposition (2) describes what evolution must be like for parsimony to have a likelihood rationale.

I have no quarrel with proposition (1). I do not think the use of cladistic parsimony is a *sui generis* requirement of rationality. If the dispute between phenetic and cladistic methods of phylogenetic inference is to be resolved, it isn't enough simply to insist that parsimony is a synonym for rationality. Instead, one must find some more neutral criterion whose legitimacy both sides can recognize. It is plausible to regard likelihood as providing that sort of benchmark.

How is X to be specified in proposition (2)? Felsenstein (1983) claims that parsimony and likelihood will coincide only if homoplasies are assumed to be very rare. He notes that this assumption is often implausible; for many real data sets, even the most parsimonious hypothesis will require a substantial number of homoplasies. As in our discussion of Sneath and Sokal's proposal, it is important to recognize that if parsimony really does *presuppose* a condition that we know to be violated, then it cannot be rational to use that method.

My own view of the matter is that Felsenstein did not establish that parsimony *requires* that homoplasies be rare (Sober 1988; Felsenstein acknowledged this point in Felsenstein and Sober 1986). What he did demonstrate was that rarity of homoplasy (when coupled with some other assumptions) *suffices* for the most parsimonious hypothesis to be the hypothesis of maximum likelihood.

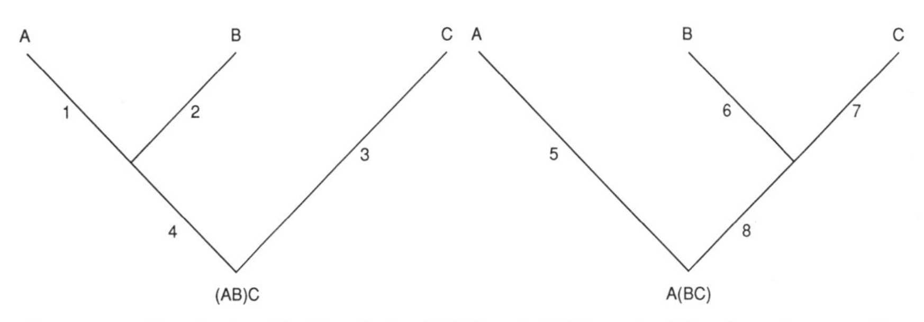

Figure 6.10 To calculate likelihoods for (AB)C and A(BC), each of the characters must be assigned *transition probabilities* for each of the branches on which it evolves.

In my opinion, what evolutionary assumptions are required for cladistic parsimony to be a legitimate principle of inference is an open question. However, I have no doubt that there must be some such set of assumptions. This flows directly from the idea that cladistic parsimony and other methods of phylogenetic inference should be evaluated from the point of view of likelihood.

It is not hard to see why adopting a likelihood framework immediately entails that evolutionary assumptions are required for parsimony to be legitimate. Consider the two trees depicted in Figure 6.10. Imagine that the $(AB)C$ hypothesis is the more parsimonious explanation of the data before us. We now wish to see whether $(AB)C$ also happens to be the hypothesis of greater likelihood: $P[\text{Data} / (AB)C] > P[\text{Data} / A(BC)]$. How are we to check whether this inequality is true?

The data set contains a number of characters. Some may have the 110 pattern; others may be distributed as 011. And there may be 001 and 100 characters in the set as well. We can decompose the likelihood of a hypothesis relative to this data set as follows: For each character, the probability of obtaining that character on a given tree will depend on the *branch transition probabilities* for the character. Once we know the likelihood of $(AB)C$ relative to each character, we need to assemble that information to tell us how likely $(AB)C$ is relative to all the data. If traits T_1 and T_2 evolve independently of each other then $P[T_1 \& T_2 / (AB)C] = P[T_1 / (AB)C]P[T_2 / (AB)C]$. However, if the traits are not independent of each other, this simple equation will not be true.

Stated in this general way, each character has its own suite of branch transition probabilities. There is no *a priori* reason why different characters must have the same set of branch transition probabilities. In addition, there is no *a priori* reason why a given character must have the same transition probabilities on every branch in a tree.

We now can write down expressions for the likelihoods of each phylogenetic hypothesis, relative to all the data we possess. It will be filled with e's and r's, each of them indexed for branches *and* for characters. These expressions will leave entirely unsettled whether the likelihood inequality stated earlier is correct. The left side of the inequality will be expressed in terms of

Box 6.3 Pattern Cladism

Although mainstream cladistics aims to reconstruct phylogenetic relationships among taxa by using the principle of parsimony, pattern cladism holds that the method of inference and the tree it singles out need not be given an evolutionary interpretation. The idea is that evolutionary ideas are *dispensable;* parsimony identifies patterns in the data of species sameness and difference that exist whether or not one thinks of these patterns as consequences of evolution (Nelson 1979; Patterson 1980, 1988; Platnick 1979).

Pattern cladism is sometimes criticized for being a throwback to phenetics. Indeed, its quest for a description of pattern that makes no assumptions about process and its skepticism about the objectivity of evolutionary ideas *are* similar to some phenetic claims (Ridley 1986). However, a point of difference remains: Whereas pheneticists classify by overall similarity, pattern cladists use parsimony, which regards only some similarities as having evidential meaning.

One very serious problem for pattern cladism is how it makes sense of parsimony without appeal to evolutionary concepts. What distinguishes ancestral from derived characters except facts about evolution? If the methodology of cladistic parsimony is inextricably connected to the idea of descent with modification in a branching process, the goal of pattern cladism cannot be achieved.

probabilities indexed for branches 1–4; the right side will be expressed in terms of probabilities indexed for branches 5–8. To render these two quantities commensurable, we have to introduce biological assumptions. A model of the evolutionary process will allow us to restate the two expressions in a common currency. Then and only then will we be able to tell which hypothesis is more likely.

It is understandable that cladists have been attracted by the idea that the principle of parsimony can be defended without making detailed assumptions about the evolutionary process. If one is ignorant of the genealogical relationships that obtain among a set of taxa, one probably will be ignorant of the processes that took place in the evolution of those taxa. If ignorance of pattern entails ignorance of process and if inference about pattern requires assumptions about process, then phylogenetic inference threatens to be a project that cannot get off the ground.

This threatened conclusion is a slippery one; before giving up in despair, we must scrutinize the connection of pattern and process more carefully. Granted, if you do not know how a set of taxa are related, then you probably will not know *some* facts about how various characters evolved. But there nonetheless may be *other* facts about the evolutionary process that are within your grasp. Consider the following example. We know that the genetic code is highly redundant. The third position in codons usually makes no difference to the amino acids that result. This suggests that changes in the third position occur mainly by random drift. This quite general line of reasoning may be defensible even though we are ignorant about the phylogenetic relationships that obtain within a given set of taxa. If so, we might be able to look at a data set of third positions in these taxa and make an inference that assumes a uniform rate model of evolution. Here, we may know *enough* about process for an inference about pattern to go forward. Knowledge of pattern cannot be wholly independent of knowledge of process, and ignorance of pattern usually entails *some* ignorance about *some* facets of process. But this is not cause for alarm.

This same line of reasoning may help us determine whether parsimony or overall similarity is a better method of inference. If each character on each branch evolves according to its own idiosyncratic but unknown rules, then nothing can be said about genealogical relationships or about the methods we should use to infer them. However, if we can say something fairly general about character evolution, or if we can identify a subset of characters that obey the same rules of change, these assumptions about process may help us resolve the conflict between parsimony and overall similarity.

Suggestions for Further Reading

Ridley (1986) is a useful introduction to the three taxonomic philosophies. Hull (1988) defends the view that species are individuals and also sets forth arguments against essentialism and pheneticism. In addition to the species concepts surveyed in this chapter, Paterson's (1985) ideas about *mate recognition systems*, Templeton's (1989) *cohesion species concept*, and the *phylogenetic species concept* (Mishler and Donoghue 1982; Dequeiroz and Donoghue 1988) are important to consider. Ereshefsky (1992) is a useful anthology of biological and philosophical work on the species concept. Finally, Sober (1988) discusses common cause and separate cause explanations in connection with Reichenbach's (1956) Principle of the Common Cause.

7

Sociobiology and the Extension of Evolutionary Theory

Sociobiology is a research program that seeks to use evolutionary theory to account for significant social, psychological, and behavioral characteristics in various species. Understood in this way, sociobiology did not begin with the publication in 1975 of E. O. Wilson's controversial book, *Sociobiology: The New Synthesis*. It is a recent phase of the long-standing effort by biologists to theorize about the evolution of behavior.

What separates sociobiology from its predecessors is its use of the vocabulary of contemporary evolutionary theory. Wilson announced that the principal problem for sociobiology is the evolution of altruism. This focus, plus the reluctance of many (but not all) sociobiologists to indulge in group selection hypotheses, is distinctive. Sociobiology is not just a research program interested in the evolution of behavior; its characteristic outlook is adaptationist, with strong emphasis on the hypothesis of *individual* adaptation.

The initial furor that arose around Wilson's book mainly concerned his last chapter, in which he applied sociobiological ideas to human mind and culture. He was criticized for producing an ideological document and charged with misusing scientific ideas to justify the political *status quo*. Sociobiology also was criticized for being unfalsifiable; sociobiologists were accused of inventing just-so stories that were not and perhaps could not be rigorously tested (Allen *et al.* 1976).

Some of these criticisms don't merit separate treatment here. My views about the charge of unfalsifiability should be clear from Chapters 2 and 5. Sociobiology, like adaptationism, is a research program; research programs do not stand or fall with the success of any one specific model.

At the same time, it is quite true that some popular formulations of sociobiological ideas have drawn grand conclusions from very slender evidence. In Chapter 5, I emphasized the importance of carefully specifying the *proposition* that an adaptationist explanation is intended to address. When a sociobiologist seeks to explain why human beings are xenophobic or aggressive or easy to indoctrinate (Wilson 1975, 1978), the first question should be: Which fact about behavior are we actually discussing? Is it the fact that human beings are *sometimes* xenophobic, that they *always* are, or that they display the trait in some circumstances but not in others? The first problem is fairly trivial, while the second is illusory; human beings are not *always* xenophobic. As is the case for adaptationist explanations generally, well-posed problems should not be too easy (Section 5.5).

Just as human sociobiology cannot be rejected on the grounds that some single sociobiological explanation is defective, the program cannot be vindicated by appealing to the simple fact that the human mind/brain is the product of evolution. What is undeniable is that theories of human behavior must be *consistent* with the facts of evolution; so, too, must they be consistent with the fact that the human body is made of matter. However, it does not follow from this that either evolutionary biology or physics can tell us anything interesting about human behavior. In Section 7.5, I will examine an idea that runs contrary to the sociobiological research program—that the human mind/brain, though a product of evolution, has given rise to behaviors that cannot be understood in purely evolutionary terms.

My own view is that there is no "magic bullet" that shows that sociobiology is and must remain bankrupt, nor any that shows that it must succeed. Any discussion of the adequacy of sociobiological models inevitably must take the models one by one and deal with details (Kitcher 1985). Obviously, a chapter in a small book like the present one does not offer space enough to carry out that task. In any event, I'm not going to try to develop any full-scale estimate of the promise of sociobiology. My interest lies in a few broad philosophical themes that have been important in the sociobiological debate.

7.1 Biological Determinism

Evolution by natural selection requires that phenotypic differences be heritable. For example, selection for running speed in a population of zebras will lead average running speed to increase only if faster-than-average parents tend to have faster-than-average offspring (Section 1.4). What could produce this correlation between parental and offspring phenotypes? The standard evolutionary assumption is that there are genetic differences among parents that account for differences in running speed. Because offspring inherit their genes from their parents, faster-than-average parents tend to have faster-than-average offspring.

This basic scenario remains unchanged when a sociobiologist seeks to explain some sophisticated behavioral characteristic by postulating that it is the result of evolution by natural selection. As mentioned earlier, Wilson (1975) suggested that human beings are xenophobic, easy to indoctrinate, and aggressive and that these behavioral traits evolved because there was selection for them. For this to be true, an ancestral population must be postulated in which there is variation for the phenotype in question. Individuals must vary in their degree of xenophobia, and those who are more xenophobic must be fitter than those who are less so. In addition, the trait must be heritable. A gene (or gene complex) for xenophobia must be postulated.

Such explanations are often criticized on the ground that there is no evidence for the existence of genes "for" the behavior in question. Even if the evidential point is correct, whether one views it as a decisive objection depends on how much of the rest of evolutionary theory one is prepared to jettison as well. Fisher (1930) constructed his model of sex ratio evolution (Box 1.3) without any evidence that there are genes for sex ratio. The same holds true for virtually all phenotypic models of evolution. Parker's (1978) optimality model of dung fly copulation time (Section 5.5) did not provide any evidence that there is a gene for copulation time, but that did not stop many evolutionary biologists from taking it seriously. It isn't that discovering the genetic mechanism would be *irrelevant* to the explanation; rather, such a discovery does not appear to be *necessary*, strictly speaking, for the explanation to merit serious consideration.

Even so, it is worth considering what it means to talk about a "gene for xenophobia" and also to consider more generally what the genetic assumptions are to which sociobiology is committed. We may begin with an assessment due to Gould (1980b, p. 91):

> There is no gene "for" such unambiguous bits of morphology as your left kneecap or your fingernail. Bodies cannot be atomized into parts, each constructed by an individual gene. Hundreds of genes contribute to the building of most body parts and their action is channeled through a kaleidoscopic series of environmental influences: embryonic and postnatal, internal and external.

"Beanbag genetics" is a pejorative label for the idea that there is a one-to-one mapping between genes and phenotypes. Gould's point is that beanbag genetics is false. But sociobiologists, in spite of the fact that they often talk about a "gene for *X*," are not committed to beanbag genetics. They can happily agree that "hundreds of genes" contribute to the phenotypes they discuss.

What does it mean to say that a gene (or complex of genes) is "for" a given phenotype? A gene *for* phenotype X presumably is a gene that *causes* phenotype X. But what does this causal claim amount to? Dawkins (1982a, p. 12) offers the following proposal:

> If, then, it were true that the possession of a *Y* chromosome had a causal influence on, say, musical ability or fondness for knitting, what would that mean? It would mean that, in some specified population and in some specified environment, an observer in possession of information about an individual's sex would be able to make a statistically more accurate prediction as to the person's musical ability than an observer ignorant of the person's sex. The emphasis is on the word "statistically," and let us throw in an "other things being equal" for good measure. The observer might be provided with some additional information, say on the person's education or upbringing, which would lead him to revise, or even reverse, his prediction based on sex. If females are statistically more likely

than males to enjoy knitting, this does not mean that all females enjoy knitting, nor even that a majority do.

Let us formulate Dawkins's idea more explicitly. If we wish to say whether being female (F) causes one to like knitting (K), we first must specify a population and an environment. So let us consider the population of human beings alive in 1993, and let the environment be the range of environments that people currently inhabit. I assume that Dawkins does not insist that the individuals considered must live in exactly the same environment since this would make it impossible to advance causal claims about the real world. Given these specifications, I take it that Dawkins's proposal is that "F causes K" means that $P(K/F) > P(K)$; this inequality is equivalent to $P(K/F) > P(K/\text{not-}F)$.

The trouble with this proposal is that it equates causation with correlation. The fact that women knit more often than men does not mean that being female is a positive causal factor for knitting. In just the same way, it may be true that drops in barometer readings are correlated with storms, but that does not mean that barometer drops cause storms (see Box 3.3).

To apply this point to the issue of what "gene for X" means, consider the fact that there are genetic differences between people living in Finland and people living in Korea. Suppose gene *a* occurs in 20 percent of the people in Finland but in 75 percent of the people in Korea. If I sample an individual at random from the combined population of these two countries and find that this individual has gene *a*, I have evidence that this person speaks Korean rather than Finnish. But from this it would be absurd to conclude that the *a* gene is a gene for speaking Korean. The *a* gene may simply be a gene for blood type; the frequencies of blood types in the two countries may be different.

There is no gene for speaking Korean. However, this does not mean that the population of Korean speakers has the same genetic profile as the population of Finnish speakers. What it means is that two people, *were they placed in exactly the same environment,* would end up speaking the same language despite whatever genetic differences they may have.

This idea can be represented schematically as follows. Suppose that everyone in the two populations has either genotype $G1$ or genotype $G2$. Suppose further that everyone is exposed to either Finnish or to Korean during early life. In principle, there are four "treatment combinations." The phenotypes that result from these four gene/environment combinations are listed as entries in the following 2 x 2 table:

		Environment Subject is exposed to	
		Finnish	Korean
Genotype	$G1$	speaks Finnish	speaks Korean
	$G2$	speaks Finnish	speaks Korean

In this example, what genotype you possess *makes no difference to the language you speak*. Of course, an individual can't speak a language without having genes; an organism won't develop at all if it has no genes. However, when we ask whether genes causally contribute to the development of some phenotype, we usually have in mind a difference between one genotype and another; the contrast between having genes of some sort and having no genes at all is not the relevant comparison.

In this 2 x 2 table, most of the individuals are either in the upper-left or the lower-right corner. People who grow up hearing Finnish tend to have genotype *G*1, and individuals who grow up hearing Korean tend to have genotype *G*2. That is, in this case, there is a *gene/environment correlation*. This correlation allows us to predict what language people speak either by knowing their environment or by knowing their genotype. Your genotype can be a good predictor of the language you speak, even though what genotype you have makes no difference in determining what language you speak.

I have just run through some of the basic ideas that biologists now use to understand the distinction between *nature* and *nurture*. It is a truism that every phenotype an organism possesses is the result of a causal process in which genetic and environmental factors interact. But given that these two sorts of causes play a role in the ontogeny of an individual, how are we to say which "contributed more" or was "more important"? Consider a phenotype like the height of a corn plant. If the plant is 6 feet tall, how are we to tell whether the plant's genes or its environment was the more important cause of its height? If the genes built 5 feet of the plant and the environment added the remaining 1 foot, we could say that the environment contributed more. But genes and environment do not work separately in this way (Lewontin 1974). How, then, are we to compare the importance of the two causal factors?

The fundamental insight of the modern understanding of this issue is that it must involve *variation in a population*. We don't ask whether genes or environment mattered more in the development of a single corn plant. Rather, we take a field of corn plants in which there is variation in height. We then ask how much of that variation can be explained by genetic variation and how much by variation in the environment.

The basic statistical idea used in this enterprise is called *the analysis of variance* (ANOVA). Again for simplicity, consider a field of corn plants in which every plant has either genotype *G*1 or *G*2 and every plant receives either one unit of water (*W*1) or two (*W*2). Suppose the four treatment cells contain the same numbers of plants and that the average heights within the cells are as follows:

		Environment	
		*W*1	*W*2
Gene	*G*1	4	5
	*G*2	2	3

In this case, shifting from *G*2 to *G*1 increases the phenotype by two units, regardless of whether the plants receive one unit of water or two. It also is true that shifting from one unit of water to two increases height by a single unit, regardless of whether the plants have genotype *G*1 or genotype *G*2. In this example, there is a *positive main effect due to genes* and a *positive main effect due to environment;* changing each makes a difference in the resulting phenotype. In addition, note that the genetic main effect is larger than the environmental main effect. Changing a plant's genes (so to speak) makes more of a difference to its height than changing its environment.

By rearranging the numbers in the above 2 x 2 table, you can construct a data set that would imply that the environmental main effect is greater than the genetic main effect. You also can describe data in which one or both of the main effects is zero. I leave these as exercises for the reader.

In the previous 2 x 2 table, influences are purely additive. Shifting from *G*2 to *G*1 means "adding" two units of height, regardless of which environment a plant inhabits; shifting from *W*1 to *W*2 means "adding" one unit of height, regardless of which genotype a plant possesses. The following data set is not additive; it involves a *gene/environment interaction*:

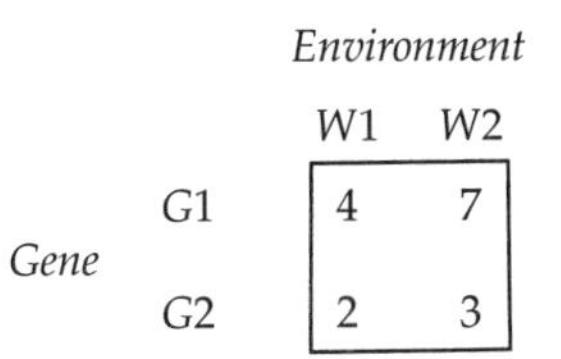

		Environment	
		*W*1	*W*2
Gene	*G*1	4	7
	*G*2	2	3

In this case, how much difference an increase in water makes depends on the plant's genotype. Symmetrically, it also is true that how much difference a change in genotype makes depends on the plant's environment. But as before, the main effects are calculated by determining how much difference *on average* a change in genes or a change in environment brings about in the resulting phenotype.

We now can clarify what it means to say that genes are more (or less) important than environment for explaining the variation of some phenotype in some population. This merely means that the genetic main effect is greater (or less) than the environmental main effect. There is a gene (or a gene complex) "for" some phenotype in a given population precisely when the variation of that phenotype possesses a genetic main effect.

One consequence of this idea is that a trait does not have to be purely "nature" (= genetic) or purely "nurture" (= environmental). *To say that genes influence some phenotype does not mean that the environment has no influence.* That genes make a difference does not mean that the environment makes no difference.

Another consequence is that it is meaningless to say that genes are more important than environment (or to advance the opposite claim) for a phenotype that does not vary. If every human being has a head, then one cannot say that genes are more important than environment in shaping this phenotype among human beings.

Even though there is no genetic main effect for the phenotype just mentioned, this does not mean that genes play no role in the ontogenetic processes in which individuals develop heads. Again, it is essential to bear in mind that "genetic main effect" has to do with whether different genes tend to produce different phenotypes. If all genes produce the same phenotype (i.e., the trait is totally *canalized*), there is no genetic main effect.

Consider another trait that is (virtually) universal within our species: Practically every human being can speak a language. Many linguists talk about an "innate language capacity," which all human beings are said to share. What could this mean, if the trait does not vary? To make sense of this idea, we must embed the human population, within which the trait is universal, in a larger population. For example, let us consider human beings together with chickens. Some individuals in this superpopulation speak a language while others do not. How do we explain this pattern of variation? Is it merely that human beings and chickens grow up in different environments? Or do genetic differences play a role?

Unfortunately, we face, at the outset, the problem of gene/environment correlation. Human beings are genetically different from chickens, but it also is true that they live in different environments. To identify the respective contributions of genes and environment, we must break this correlation, or, since ethical considerations prevent us from doing this, we must try to figure out what would happen if the correlation were broken. Just as in the example about Korean and Finnish, we need to fill in all four cells in the following 2 x 2 table:

	Environment	
	Exposed to a human language	Not exposed to a human language
Human genes	Yes	No
Chicken genes	No	No

The four entries describe whether the individual will speak a language. In this example, the contributions of genes and environment are entirely symmetrical. Having the right genes is essential, but so, too, is living in the right environment.

Apportioning causal responsibility between genes and environment depends on the set of genes and the range of environments considered. Consider, for example, the genetic disease known as PKU syndrome (phenylketonuria). Individuals with two copies of the recessive gene (call it "*p*") cannot digest phenylalanine. If their diet contains phenylalanine, they will develop a severe retardation. However, *pp* homozygotes will develop quite normally if their diet is carefully controlled.

Let us consider PKU syndrome both before and after these facts about its dietary control were discovered. Before the discovery, pretty much everyone ate diets that contained phenylalanine. In this case, the explanation of why some individuals ended up with PKU syndrome while others did not was entirely genetic. However, once the diet of *pp* homozygotes was restricted, the causal profile of PKU syndrome changed. Today, it is true that both genes *and* environment make a difference; the syndrome now is no more genetic than it is environmental (Burian 1983).

A simpler example illustrates the same point. Suppose that a set of genetically different corn plants are raised in the same environment; differences in height then must be due solely to genetic differences. But if you take that same set of corn plants and raise them in a variety of environments, the environmental main effect now may be nonzero—indeed, it may even be larger than the genetic main effect. *Whether a phenotype is mainly genetic is not an intrinsic feature that it has but is relative to a range of environments* (Lewontin 1974).

The question "Do genes matter more than environment?" is meaningless. This query must be relativized to a phenotype. Which language you speak is determined by your environment, but your eye color is determined by your genes. In addition to specifying the phenotype in question, one also must fix the range of genes and environments one wishes to consider. A trait can be mainly genetic in one range of environments but fail to be so in another.

Given this account of what it means to talk about a gene (or genes) for *X* (where *X* is some phenotype), I now want to consider what sociobiology presupposes about the issue of genetic causation. Sociobiologists sometimes discuss traits that they take to be universal (or nearly so) within the species of interest. At other times, they discuss traits that show within-species variation. Let us take these two cases in turn.

I have already mentioned that evolution by natural selection requires that the evolving trait be heritable (or "heritable in the narrow sense," meaning that there is a genetic main effect in explaining the phenotypic variation). We now must see that the heritability of an evolving trait itself evolves. The fact that a trait must be heritable while it is evolving does not mean that it must remain heritable after it has finished evolving.

Consider a simplified scenario for the evolution of the opposable thumb. There was an ancestral population in which some individuals had opposable thumbs while others did not, and this phenotypic difference reflected genetic differences between the two classes of individuals. Selection then caused opposable thumbs to increase in frequency; eventually, the trait went to fixation. At this point, the gene(s) for an opposable thumb also became fixed. Since there now is no phenotypic variation in the population, there is no genetic main effect. Hence, the heritability is zero. Just as natural selection typically destroys variation (Sections 1.4 and 5.2), it also destroys heritability.

Of course, it is an oversimplification to say that there is absolutely no variation for the phenotype in present-day populations. Thirty years ago, a number of women took the drug thalidomide, not knowing that this would cause their children to be born without arms. In addition, people sometimes lose their thumbs in industrial accidents. Examples such as these show that it could easily be true that present-day variation for the phenotype of having an opposable thumb is mainly environmental and nongenetic. So when a sociobiologist posits a gene for *X* by way of explaining why the *X* phenotype

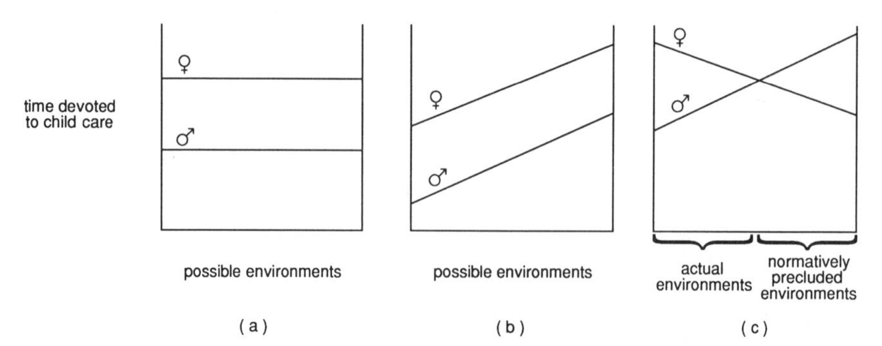

Figure 7.1 Three possible relationships between the average amount of time that women and men devote to parental care.

evolved to fixation, the genetic main effect that this demands must have existed ancestrally. It need not exist today.

Although such explanations involve no commitment to the existence of present-day genetic variation, suppose we found that such genetic variation exists. Does this automatically lend credence to a selective explanation? Here, we must be careful. If a trait is said to have evolved because of the strong selective advantage it provides, we should be puzzled as to why genetic variation for the trait still exists. It is not uncommon for sociobiologists to simultaneously say that a phenotype (like intelligence) was shaped by selection and to claim that the trait now has an important genetic component. Far from fitting together harmoniously, there is a dissonance between these two claims that we must learn to hear.

Now let us consider traits that sociobiologists think presently show within-species variation. One prominent example is the matter of behavioral differences between the sexes. Sociobiologists sometimes suggest that men are more promiscuous than women (and that women are more "coy" than men) and that evolutionary theory explains why. They also have commented on differences in patterns of child care, discussing why women stay home with the kids more than men do.

Let us focus on parental care. Suppose that women, on average, take care of their children more than men do in each of the various environments that human beings have inhabited to date. This difference between the sexes might obey three different patterns, depicted in Figure 7.1.

In part (a) of the figure, men and women differ in the average amount of time each spends on parental care. Note that the average amount of time spent by each sex is not affected by the environment. In part (b), the amount of time *is* influenced by changes in the environment, although the difference between the sexes is not.

Part (a) represents a stronger form of "biological determinism" than part (b). It is a curious terminological fact that "biological determinism" is so often used to mean *genetic* determinism (as if environmental causes like nutrition were not "biological"). Part (a) says that the absolute value *and* the relation between the sexes cannot be modified by the environment; part (b) says that changing the environment can modify the absolute value for each sex but not the relation between the sexes.

Sociobiology is not committed to the ideas depicted in either part (a) or part (b). An evolutionary explanation of behavioral differences between the sexes does not have to maintain that there is *no possible environment* in which this difference would be erased or reversed. Sociobiologists often maintain that it would be very hard to eliminate certain behavioral differences between the sexes (Wilson 1975). For example, it might be necessary to completely overhaul the pattern of child care that boys and girls experience. Perhaps biological parents would have to be replaced by child-rearing experts who are trained by the state to behave in certain ways. In this radically altered environment, girls and boys might grow up to be parents who provide equal amounts of parental care. A sociobiologist might argue that this arrangement, though not impossible, would be undesirable. The new arrangement would require sacrificing values that many hold dear (Kitcher 1985).

This third possibility is depicted in Figure 7.1(c). In this arrangement, the environment affects not just the absolute amount of child care but whether women provide more of it than men. Although part (a) and part (b) represent versions of biological (i.e., genetic) determinism, part (c) cannot be interpreted in this way.

In all three figures, the behavioral difference between the sexes within the range of *actual* environments is said to have a nonenvironmental cause. If the x axis represents all environmental causes, then the genetic difference between men and women (presumably, the fact that women are usually *XX* and men are usually *XY*) is said to have explanatory relevance.

As became clear in discussing Dawkins's knitting example, it is important not to be misled by gene-environment correlations. The fact that *XX* individuals, on average, provide more child care than *XY* individuals does not, by itself, entail that *XX* is a genetic configuration that codes for greater child care. Only if we control for environmental causes and *still* find that there is a genetic main effect can we conclude that this behavioral difference between the sexes has a genetic cause.

Sociobiologists generally have favored the hypothesis that important behavioral differences between the sexes have a significant genetic component. Selection has favored different behaviors in the two sexes. *Within women,* selection has favored one set of behaviors; *within men,* it has favored a different set. Of course, this hypothesis does not exclude what is obviously true—that some men provide more parental care than others and that there is variation among women for the trait as well. The hypothesis attempts to account for variation *between* the sexes, not *within* them. Variation *within* the sexes may be mainly environmental, but variation *between* the sexes, so the selectionist explanation implies, will have a significant genetic component.

Although many sociobiologists are inclined to explain this pattern of variation in genetic terms, it is not an inevitable commitment of sociobiological theorizing that all within-species variation must be explained in this way. A useful example of why this isn't intrinsic to the research program is provided by the work of Richard Alexander (1979). Alexander is interested in explaining within-species variation. For example, he addresses the question of why some societies but not others follow the kinship system known as the avunculate. In this arrangement, men provide more care for their sisters' children than for the children of their spouses. Alexander suggests that this kinship system occurs in societies in which men are very uncertain about paternity. If women are sufficiently promiscuous, a man will have more genes in common with his sister's children than with the children of his wife. Thus, a man within such a society maximizes his reproductive success by helping his sister's children, rather than helping his wife's.

I don't want to address the empirical issues of whether this explanation is correct. My point is that Alexander is not asserting that societies that follow the avunculate differ genetically from societies that do not. According to Alexander, human beings have a genetic endowment that allows them enormous behavioral flexibility. The human genotype has evolved so that individuals adjust their behaviors in a way that maximizes fitness. People in different societies behave differently not because they are genetically different but because they live in different environments. The avunculate maximizes fitness in some environments but not in others.

In a curious way, Alexander is a "radical environmentalist" with respect to within-species variation. Far from wishing to explain behavioral differences as "in our genes," he holds that behavioral variation is to be explained environmentally. This is about as far from a commitment to biological (i.e., genetic) determinism as one can get.

7.2 Does Sociobiology Have an Ideological Function?

Critics have seen sociobiology as the latest installment in a long line of biological ideas, stretching from the social Darwinism popular at the end of the nineteenth century through the IQ testing movement around the period of World War I to Nazi "racial biology," and to the debate about race and IQ in the 1960s (Chorover 1980; Lewontin, Rose, and Kamin 1984). Sociobiology, like its predecessors in this lineage, is charged with being ideological.

What might this charge of "ideology" mean? Several distinctions are needed. First, it might be claimed that individual authors or the people who determine which ideas are disseminated or the general readership of these views are motivated by ideological considerations. Second, there is the issue of *how much* of a role ideological considerations play in this three-step process of creation, dissemination, and acceptance. An extreme version of the ideology thesis might claim that there is not a shred of scientific evidence in support of sociobiological ideas, so the ideas are formulated, disseminated, and accepted for entirely nonscientific reasons. A less extreme thesis would be that the degree of conviction that people have with respect to these ideas far outruns the evidence actually at hand; what should be regarded as speculation gets interpreted as established truth.

In all its guises, the ideology thesis is a thesis of *bias*. It claims that something influences the production/dissemination/acceptance process besides

Box 7.1 The Ought-Implies-Can Principle

Sociobiology has been criticized for defending the political *status quo*. If sociobiology entailed a strong thesis of biological determinism, the charge would make sense. If existing inequalities between the sexes or among the races or among social classes were biologically unalterable, then this fact would undercut criticisms of existing social arrangements. It would be hopelessly utopian to criticize society for arrangements that cannot be changed.

The argument just stated makes use of the *Ought-Implies-Can Principle*: *If a person ought to do X, then it must be possible for the person to do X*. If you *cannot* save a drowning person (e.g., because you cannot swim or have no access to a life preserver), then it is false that you *ought* to save that person: You cannot be criticized for not doing the impossible. Likewise, if our biology makes it impossible for us to eliminate certain inequalities, then it is false that we ought to eliminate those inequalities.

If the Ought-Implies-Can Principle is correct, then scientific results can entail that various ought-statements are false. Does this entailment relation contradict Hume's thesis (Section 7.4) about the relation of is-statements and ought-statements?

evidence; that extra something is the goal of advancing a political agenda. The ideology thesis does not entail that individual sociobiologists have been biased—the mindset of individual authors pertains to the production of sociobiological ideas, not to their subsequent dissemination or acceptance by a larger community. Suppose the mass media were biased in favor of publicizing scientific ideas that could be interpreted as justifying the political *status quo*. If sociobiological ideas were disseminated because they could be so interpreted, then sociobiology would perform an ideological function *even if no individual sociobiologist departed from reasonable norms of scientific objectivity*. Perfectly objective scientific findings can be put to ideological use.

I am here putting to work an idea explored in Section 3.7 concerning what it means to ascribe a function to something. What does it mean to say that the heart has the function of pumping blood but not of making noise? One suggestion is that the functional statement makes a claim about why the heart is there: Hearts persist because they pump blood, not because they make noise. In ascribing an ideological function to sociobiology, critics are making a claim about why such ideas persist.

Understood in this way, this functional claim is not obviously true. An empirical argument is needed to show that some part of the production/dissemination/acceptance process is biased and that the bias is due to the goal of advancing some political agenda. Glib statements about the "bias" of the mass media notwithstanding, it is no small task to muster evidence for claims of this sort.

Take a quite different and possibly simpler functional explanation of the persistence of an idea. Malinowski (1922) wanted to account for why South Sea Islanders have elaborate rituals surrounding deep-sea fishing but none connected with fishing in fresh water. His explanation was that deep-sea fishing is far more dangerous than fishing in fresh water and that rituals evolved in connection with the former because they reduce fear.

To test Malinowski's conjecture, at least two hypotheses would have to be investigated. The first—that the rituals actually do reduce fear—might be investigated by an experiment in which some individuals are exposed to the rituals while others are not. We would like to know whether the first group is less fearful than the second. If deep-sea rituals really do reduce fear, the next question would be whether the rituals persist *because* they have this effect. In this connection, we would like to know if the rituals would persist even if they did not reduce fear. It is possible, after all, that rituals promote group solidarity and persist for this reason, quite apart from their effect on fear. If other rituals persist for reasons having nothing to do with fear reduction, this makes it less than transparent that deep-sea fishing rituals persist because they reduce fear. Perhaps the experiment to consider here would be to make deep-sea fishing quite safe. Would the rituals then wither away?

To document the claim that sociobiology has an ideological function, a similar pair of questions must be posed. Do sociobiological ideas really convince people that existing inequalities are legitimate and inevitable? This question is not settled simply by looking at what sociobiologists *say*. The issue is what impact various lectures, books, and articles have on their audience. Do people who read sociobiology accept the political *status quo* more than the members of some control group do? This is not obvious, but it may be true.

If sociobiological ideas do have this consequence, the second step would be to determine whether sociobiological ideas persist *because* they have this effect. Would such theorizing continue if it ceased to be understood as justifying the *status quo*? As in the case of deep-sea fishing rituals, the answer is not obvious. It may be that sociobiological theorizing is driven by a dynamic of scientific investigation that would propel the research program even if no one interpreted it as having political implications—after all, the evolution of behavior is an enticing problem area for biologists. Perhaps some sociobiological ideas persist for purely scientific reasons.

I said before that the question of whether sociobiology has an ideological function may be more complicated than the question of whether deep-sea fishing rituals have the function of reducing fear. One reason is that sociobiology is not a single idea; it is a web of various ideas, loosely connected with each other but elaborately connected with diverse elements in the rest of biology. It is possible that some themes in sociobiology persist for ideological reasons while others stay afloat for wholly scientific reasons. Just as there is no simple and global answer to the question of whether sociobiology is *true*, so there is no simple and global answer to the question of whether sociobiology functions to justify existing political arrangements. A serious investigation of either issue must proceed piecemeal.

Critics charge that sociobiology is ideology, not science. Sociobiologists protest that their own motives are scientific and that it is the critics themselves (some of them Marxists) who are ideologically motivated. All this mudslinging aside, there is an issue here about the sociology of ideas that is worth considering seriously. It is no great shock to the scientific temperament to consider the possibility that *religious* ideas may persist for reasons having nothing to do with their truth. South Sea Islanders perform rituals to appease the gods; the rituals persist but not because there are gods who answer the Islanders' prayers. When this style of explanation is applied to the content of science, it is more difficult for scientists to approach it objectively. Yet, it is a possibility deserving of scientific scrutiny that some scientific ideas persist for reasons other than their evidential warrant. One should not accept this suggestion glibly, but neither should it be dismissed out of hand.

7.3 Anthropomorphism Versus Linguistic Puritanism

The next criticism of sociobiology I want to consider is aimed at suggestions like the following one, which was put forward by David Barash in his book *The Whispering Within* (1979, pp. 54, 55):

> Some people may bridle at the notion of rape in animals, but the term seems entirely appropriate when we examine what happens. Among ducks, for example, pairs typically form early in the breeding system, and the two mates engage in elaborate and predictable exchanges of behavior. When this rite finally culminates in mounting, both male and female are clearly in agreement. But sometimes strange males surprise a mated female and attempt to force an immediate copulation, without engaging in any of the normal courtship ritual and despite her obvious and vigorous protest. If that's not rape, it is certainly very much like it.
>
> Rape in humans is by no means as simple, influenced as it is by an extremely complex overlay of cultural attitudes. Nevertheless, mallard rape and bluebird adultery may have a degree of relevance to human behavior. Perhaps human rapists, in their own criminally misguided way, are doing the best they can to maximize their fitness. If so, they are not that different from the sexually excluded bachelor mallards. Another point: Whether they like to admit it or not, many human males are stimulated by the idea of rape. This does not make them rapists, but it does give them something else in common with mallards. And another point: During the India-Pakistan war over Bangladesh, many thousands of Hindu women were raped by Pakistani soldiers. A major problem that these women faced was rejection by husband and family. A cultural pattern, of course, but one coinciding clearly with biology.

Critics maintain that three errors occur in this and similar sociobiological accounts. First, there is *anthropomorphism:* A term ("rape") designed for application to human beings is extended to other species. Second, there is uncritical *adaptationism:* An explanation is invented for the mallard behavior that is not well supported by evidence. Third, the adaptationist explanation of the trait in ducks is read back into our own species.

The middle criticism I will not address here; I want to focus on the first and third. Why is it a mistake to think that a human behavior is "the same" as a trait found in some nonhuman species? And why should it be a mistake to think that the explanation of a trait found in a nonhuman species also applies to the human case?

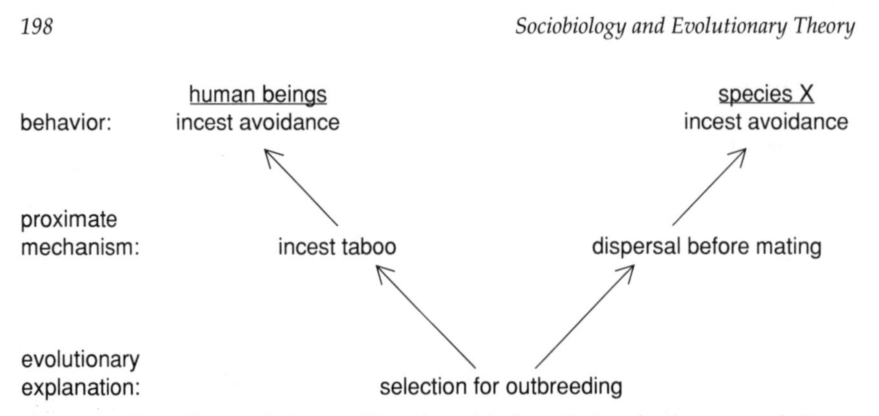

Figure 7.2 Even if human beings avoid mating with close relatives for the same *evolutionary* reason that the members of species *X* do, it does not follow that the behavior is under the control of the same *proximate mechanism* in the two species.

To address these questions, I'll shift to another example, which is one of sociobiology's favorites—the existence of incest avoidance. Virtually all human cultures restrict or prohibit individuals from reproducing with close relatives. True, the pharaohs of ancient Egypt engaged in brother/sister marriages, but this is very much the exception rather than the rule. The sociobiological explanation of incest avoidance is that inbreeding increases the probability that offspring will have two copies of deleterious recessive genes. In consequence, natural selection has caused us to outbreed.

The sociobiological explanation just sketched applies to humans and nonhumans alike. Yet, the explanation does not deny that human beings are unique; human beings avoid the behavior in part because they have an incest *taboo*. A taboo is a socially institutionalized system of beliefs and values. Human beings, unlike other organisms, avoid incest (to the extent they do) because of the beliefs and values that they have.

So the sociobiological account of incest avoidance says that human beings are unique in one respect but not in another. From an evolutionary point of view, we avoid incest for the same reason that other species do. However, the proximate mechanism that leads individual human beings to avoid inbreeding differs from the one that leads members of other species to do so.

This idea is represented in Figure 7.2. Consider some insect species *X* that has little inbreeding because individuals disperse from the nest before mating at random. Although human beings and species *X* avoid inbreeding for the same evolutionary reason, the proximate mechanisms are different. Here, we are using Mayr's (1961) distinction between proximate and ultimate explanation (Section 1.2). When sociobiologists explain incest avoidance in human beings by appealing to the selective advantage of outbreeding, they are not describing what goes on in the minds of human beings. They are attempting to describe evolutionary causes, not psychological (proximate) mechanisms.

It follows that the question "Why do human beings avoid incest?" can be addressed at two levels of analysis. One might try to answer it by discussing

Box 7.2 Incest

The sociobiological explanation of incest avoidance predicts that incest will be rare. But how rare is it? That depends on how "incest" is defined. If it is defined narrowly to mean reproduction between close relatives, we get one answer, but if it is defined more broadly to mean sexual contact between close relatives, we get another. Given how common sexual abuse of children is, perhaps the problem to address is not why incest is so rare but why it is so common.

Even if we opt for the narrower definition of incest, the question remains of how rare incest must be for the sociobiological explanation to be accepted. Can we shrug off a nonzero rate of reproduction among close relatives as compatible with the evolutionary model? How much incest would there have to be for us to conclude that the evolutionary explanation has been disconfirmed? Presumably, there is no threshold value.

Perhaps the comparative approach (Section 5.5) is more promising. Let us discover how much reproduction among close relatives there is within a variety of species, our own included. Then let us address the problem of explaining this pattern of variation. The simple idea that inbreeding is deleterious will not be sufficient. In some species, brother/sister mating is routine; in others, it is rare. If we jettison the simple question "Why do human beings avoid incest?" and substitute a comparative problem in its stead, our task becomes more difficult but also more interesting.

human psychology, or one might try to provide an evolutionary explanation. This is not to endorse what sociobiologists say about incest. My point is that *the psychological and the evolutionary answers are not in conflict.*

This idea has considerable relevance in evaluating Barash's explanation of rape in human beings. It is sometimes claimed that (human) rape should be regarded as an act of violence, not as a sexual act. The thought here is that rapists want to exercise power over their victims; it isn't sexual desire that drives the rapist but the desire to intimidate, humiliate, and punish (Brownmiller 1975). I will not try to assess whether this is a correct hypothesis about the psychology of rape. The point to recognize is that it is quite compatible with the sociobiological explanation. According to Barash, rape evolved because of the reproductive advantage it provided for rapists; this says nothing, *per se,* about the psychological motives that make rapists do what they do.

This observation does not resolve our initial question about whether rape should be defined broadly so that it applies to nonhuman organisms or narrowly so that it is uniquely human. The distinction of proximate from evolutionary explanations shows that whichever way we define the behavior, both psychological and evolutionary questions can be posed about why rape occurs. Let us now turn to the issue of broad versus narrow definition.

Although choices of terminology may appear arbitrary, they often reflect assumptions about how research problems should be organized. If rape is defined as "forced reproduction," the term gathers together some behaviors but not others. Likewise, if it is defined as "an act of sexual violence motivated by the desire to exercise power," some behaviors but not others are gathered together.

According to ordinary usage, rape can occur without reproduction. It can involve oral or anal heterosexual acts and also coerced homosexual activity. The definition of rape as "forced reproduction," however, will not count such acts as rapes. Barash wants to find a common explanation of forced reproductive activity in humans and mallards, but this way of formulating the problem involves no special obligation to have the same explanation also cover nonreproductive behavior. By the same token, those who define rape as "an act of sexual violence motivated by the desire to exercise power" will want to provide a common explanation for coerced heterosexual intercourse and the sexual acts just mentioned. But they will feel no special obligation to have an explanation of rape also apply to the behavior of mallards. Each choice of terminology brings one set of acts to the foreground, demanding a common explanation and consigns another set of acts to the background, comprising an unrelated explanatory problem.

These conflicting taxonomies of behavior reflect the difference between what "sex" means to evolutionary biologists and what it means in ordinary language. For the evolutionary biologist, sex is a distinctive mode of reproduction found in many plants and animals. In ordinary language, sexual activity includes but is not limited to reproductive activity. An important part of what makes an act "sexual," in this vernacular sense, is the intentions of the actors.

It is by no means obvious that all or even most sexual activity, in the vernacular sense of that term, should be understood in terms of natural selection. Human mind and culture have given sexuality an amazingly complicated elaboration. To understand sexual behavior in terms of its relation to reproduction may be no more promising than understanding food customs in terms of their contribution to nutrition. Just as there is more to eating than survival, so there is more to sex than reproduction.

When Barash suggests that "rape" is a trait found in both human beings and in mallard ducks, he is saying that the explanation of the trait in one species has something significant in common with its occurrence in the other. We can use the concepts of *homology* and *homoplasy,* discussed in Chapter 6, to map out some of the options. Homologies, recall, are similarities inherited unmodified from a common ancestor. Homoplasies are similarities that evolved independently in the two lineages. Consider two species *S*1 and *S*2 that both exhibit some trait *T*. Figure 7.3 depicts three possible explanations of this similarity.

To illustrate the difference between what I am calling functionally similar and functionally dissimilar homoplasies, consider two examples. Wings in birds and wings in insects are homoplasies, but they evolved for very similar functional reasons. In both lineages, wings evolved because there was selection for flying and wings facilitate flight. Consider, in contrast, the fact that lizards and ferns are both green. This similar coloration is not inherited from a common ancestor; in addition, the reason the color evolved in one lineage has

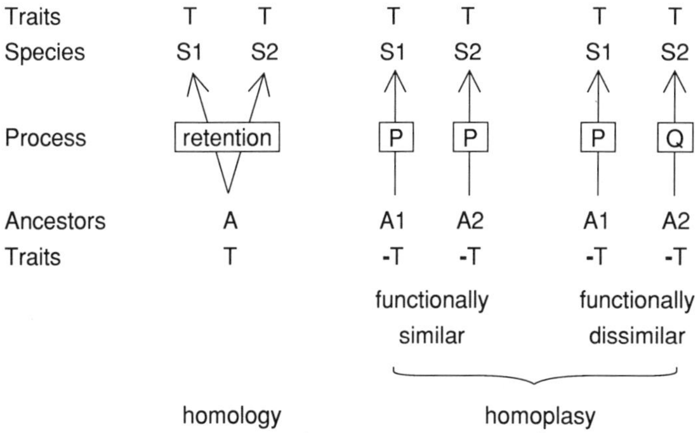

Traits	T	T	T	T	T	T
Species	S1	S2	S1	S2	S1	S2
Process	retention		P	P	P	Q
Ancestors	A		A1	A2	A1	A2
Traits	T		-T	-T	-T	-T
			functionally similar		functionally dissimilar	
	homology		homoplasy			

Figure 7.3 Three scenarios for the evolution of a similarity between species *S1* and *S2*. They can share trait *T* as a homology, as a functionally similar homoplasy, or as a functionally dissimilar homoplasy.

nothing functionally in common with the reason it evolved in the other. The occurrence of wings in birds and insects is a *functionally similar homoplasy*; the occurrence of greenness in lizards and ferns is a *functionally dissimilar homoplasy*.

When Barash applies the term "rape" to both human beings and mallard ducks, his point in using the common term is to suggest that the behaviors are either homologous *or* functionally similar. What is excluded by this sociobiological idea is that the apparent similarity between the behaviors is superficial and ultimately misleading.

Sociobiologists and their critics will agree that "greenness" in lizards and ferns is a functionally dissimilar homoplasy. Other traits are more controversial. "Rape" is the example I have discussed so far, but the same question arises in connection with other sociobiological explanations. For example, Wilson (1978) suggests that homosexuality in human beings evolved for the same reason that sterile castes evolved in the social insects. Sterile workers help their siblings to reproduce. The suggestion is that homosexuals do not reproduce but indirectly lever their genes (including "genes for homosexuality") into the next generation by helping heterosexual family members with child care.

The term "homosexuality" requires clarification. Once this is supplied, it is important to see what testable consequences follow from Wilson's proposal. For example, does his hypothesis predict that every family should contain homosexual offspring (just as every nest in a species of social insects contains sterile workers)? In addition, the hypothesis seems to predict that species in which there is more parental care should contain more "homosexual activity" than species ir which there is less.

When critics of sociobiology object to the application of terms like "rape" to nonhuman organisms, sociobiologists often reply that the critics are trying to limit terminology for no good reason. After all, "selection" used to be a term that implied conscious choice, but Darwin saw the point of using the term in a "larger and metaphorical sense." Critics charge sociobiology with anthropomorphism; sociobiologists charge their critics with linguistic puritanism. These charges and countercharges easily suggest that the dispute involved here is not substantive. After all, it is up to us how we define our terms, and surely there is no serious issue about which definition is "really" correct (see the discussion of definitions in Box 1.1). But to dismiss the dispute about terminology in this way is to miss the substantive question that underlies it. The real problem is homology and functionally similar homoplasy, on the one hand, versus functionally dissimilar homoplasy, on the other.

7.4 Ethics

Sociobiologists have addressed two very different classes of questions about ethics. The first concerns why we believe the ethical statements that we do. If there are ethical beliefs that are held in all human cultures, then evolution may help to explain why these beliefs are universal. And values that vary from culture to culture also have been addressed by sociobiologists, for example via the hypothesis, favored by Alexander (1979, 1987), that human beings adjust their behavior to maximize fitness. Neither of these enterprises can be rejected *a priori;* everything depends on the extent to which specific hypotheses are supported by specific data.

In addressing the problem of explaining morality, it is important to break the phenomenon we call "morality" into pieces. Rather than asking whether "morality" is the product of natural selection, we should focus on some specific *proposition* about morality. Even if evolution helps explain why human societies possess moral codes, it is a separate question whether evolution helps explain the specific contents of those codes (Ayala 1987). Perhaps there is a simple evolutionary explanation for why no society demands universal infanticide. On the other hand, it is not at all clear that evolutionary theory helps explain why opinion about the morality of slavery changed so dramatically in Europe during the nineteenth century. Rather than look for some sweeping global "explanation of morality," it is better to proceed piecemeal.

The second kind of question about ethics that sociobiologists have addressed is of an entirely different sort. Sometimes, the claim is advanced that evolutionary theory can tell us what our ethical obligations are. At other times, it is argued that the facts of evolution show that ethics is an illusion: although evolution leads us to *believe* that there is a difference between right and wrong, there really is no such thing (Ruse and Wilson 1986; Ruse 1986). In both instances, evolutionary theory is thought to tell us which ethical statements (if any) are true. It is this kind of project that I want to discuss now.

A common but by no means universal opinion among scientists is that *all facts are scientific facts.* Since ethical statements—statements about what is right or wrong—are not part of the subject matter of any science, it follows that there are no ethical facts. The idea is that in science, there are opinions *and* facts; in ethics, there is only opinion.

Let us say that a statement describes something subjective if its truth depends on what some subject believes; a statement describes something objective, on the other hand, if its truth or falsity is independent of what anyone believes. "People believe that the Rockies are in North America" describes something subjective. "The Rockies are in North America," on the other hand, describes something objective. When people study geography, there is both a subjective and an objective side; there are opinions about geography, but in addition, there are objective geographical facts.

Many people now believe that slavery is wrong. This statement describes something subjective. Is there, in addition to this widespread belief, a fact regarding the issue of whether slavery really is wrong? *Ethical subjectivism,* as I will use the term, maintains that there are no objective facts in ethics. In ethics, there is opinion and nothing else.

According to subjectivism, neither of the following statements is true:

Murder is always wrong.

Murder is sometimes permissible.

Naively, it might seem that one or the other of these statements must be true. Subjectivists disagree. According to them, no ethical statement is true. Hume (1739, pp. 468–469) can be viewed as endorsing subjectivism in the following passage from his *Treatise of Human Nature*:

> Morality [does not consist] in any matter of fact, which can be discover'd by the understanding. ... Take any action allow'd to be vicious: Wilful murder, for instance. Examine it in all lights, and see if you can find that matter of fact, or real existence, which you call vice. In whichever way to you take it, you find only certain passions, motives, volitions, and thoughts. There is no other matter of fact in the case. The vice entirely escapes you, as long as you consider the object. You never find it, til you turn your reflexion into your own breast, and find a sentiment of disapprobation, which arises in you, towards this action. ... It lies in yourself, not in the object.

For Hume, the whole of ethics is to be found in the subject's feelings about murder; there is not, in addition, an objective fact about whether murder really is wrong.

Ethical realism conflicts with ethical subjectivism. Realism says that in ethics, there are facts as well as opinions. Besides the way willful murder may make you feel, there is, in addition, the question of whether the action really is wrong. Realism does not maintain that it is always obvious which actions are right and which are wrong—realists know that uncertainty and disagreement surround many ethical issues. However, for the realist, there are truths in ethics that are independent of anyone's opinion.

This book is not the place to attempt a full treatment of the dispute between subjectivism and realism. However, I do want to discuss two arguments that attempt to show that ethical subjectivism is true. I will suggest that neither of these arguments is convincing.

The first has its provenance in a logical distinction that Hume drew between what I will call *is-statements* and *ought-statements*. An *is-statement* describes what is the case without making any moral judgment about whether this situation is good or bad. An *ought-statement,* on the other hand, makes a moral judgment about the moral characteristics (rightness, wrongness, etc.) that some action or class of actions has. For example, "Thousands of people are killed by handguns every year in the United States" is an is-statement; "It is wrong that handguns are unregulated" is an ought-statement.

Hume defended the thesis that ought-statements cannot be deduced from exclusively is-statements. For example, he would regard the following argument as deductively invalid:

Torturing people for fun causes great suffering.

Torturing people for fun is wrong.

The conclusion does not follow deductively from the premisses. However, if we supply an additional premiss, the argument can be made deductively valid:

Torturing people for fun causes great suffering.
It is wrong to cause great suffering.

Torturing people for fun is wrong.

Notice that this second argument, unlike the first, has an ought-statement as one of its premisses. Hume's thesis says that *a deductively valid argument for an ought-conclusion must have at least one ought-premiss.*

The term "naturalistic fallacy" is sometimes applied to any attempt to deduce ought-statements from exclusively is-premisses. The terminology is a bit misleading: It was G. E. Moore in *Principia Ethica* (1903) who invented the idea of a "naturalistic fallacy," and his idea differs from the one just described. However, since most people discussing evolutionary ethics tend to use Moore's label to name Hume's insight, I will follow them here. Hume's thesis is that the naturalistic fallacy is, indeed, a fallacy: You can't deduce an *ought* from an *is.*

Hume's thesis, by itself, does not entail subjectivism. However, it plays a role in the following argument for subjectivism:

(S1) (1) Ought-statements cannot be deduced validly from exclusively is-premisses.
(2) If ought-statements cannot be deduced validly from exclusively is-premisses, then no ought-statements are true.

No ought-statements are true.

Premiss (1) is Hume's thesis. Premiss (2), which is needed to reach the subjectivist conclusion, is a *reductionist assumption*. It says that for an ought-statement to be true, it must reduce to (be deducible from) exclusively is-premisses.

My doubts about argument (S1) center on premiss (2). Why should the fact that ethics cannot be deduced from purely *is*-propositions show that no ethical statements are true? Why can't ethical statements be true though irreducible? It is worth noting that Hume's thesis concerns *deductive* arguments. Theories about unobservable entities cannot be deduced from premisses that are strictly about observables, but this provides no reason to think that theories about unobservables are always untrue.

There is another lesson that we can extract from Hume's thesis. When biological premisses are used to argue for some ethical conclusion, there must be ethical assumptions in the background. When these assumptions are flushed into the open, the arguments sometimes look quite implausible. For example, Wilson (1980, p. 69) points out that homosexual behavior is found in nature and thus is as "fully 'natural' as heterosexual behavior." Can we conclude from this that there is nothing immoral about homosexuality? We can, provided that we are prepared to append some further premiss of an ethical sort (for example, that all "natural" behaviors are morally permissible). More recently, the same ethical conclusion has been said to flow from the hypothesis that there may be a genetic component to homosexuality. Although I am fully in sympathy with the ethical conclusion, I think these arguments on its behalf are frail. Surely there are traits found in nature (and traits that have a genetic component) that are morally objectionable. Homophobia is a bad thing, but these are bad arguments against it.

I now want to consider a second argument for ethical subjectivism. It asserts that ethical beliefs cannot be true because the beliefs we have about right and wrong are merely the product of evolution. An alternative formulation of this idea would be that subjectivism must be true because our ethical views are produced by the socialization we experience in early life. These two ideas may be combined as follows:

(S2) We believe the ethical statements we do because of our evolution and because of facts about our socialization.

———————

No ethical statement is true.

Philosophers are often quick to criticize such arguments for committing the so-called *genetic fallacy*. "Genetic" here has nothing to do with chromosomes; rather, a genetic argument describes the genesis (origin) of a belief and attempts to extract some conclusion about the belief's truth or plausibility.

The dim view that many philosophers take of genetic arguments reflects a standard philosophical distinction between the *context of discovery* and the *context of justification*. This distinction, emphasized by the logician Gottlob Frege, was widely embraced by the positivists. Hempel (1965b) tells the story of the chemist Kekulé who worked on the problem of determining the structure of benzene. After a long day at the lab, he found himself gazing wearily at a fire. He hallucinated a pair of whirling snakes, which grabbed each other's tails and formed a circle. Kekulé, in a flash of creative insight, came up with the idea of the benzene ring.

The fact that Kekulé arrived at this idea while hallucinating has nothing to do with whether benzene really has a ring structure. It is for psychologists to describe the context of discovery—the idiosyncratic psychological processes that led Kekulé to his insight. After he came up with this idea, he was able to do experiments and muster evidence. This latter set of considerations concerns the logic of justification.

I agree that one can't *deduce* that Kekulé's hypothesis was true or false just from the fact that the idea first occurred to him in a dream. But it is a mistake to overinterpret this point. I suggest that there can be perfectly reasonable genetic arguments. These will be *non*deductive in form.

Consider an example. Suppose I walk into my introduction to philosophy class one day with the idea that I will decide how many people are in the room by drawing a slip of paper from an urn. In the urn are a hundred such slips, each with a different number written on it. I reach in the urn, draw a slip that says "78," and announce that I believe that exactly 78 people are present.

Surely it is reasonable to conclude that my belief is probably incorrect. This conclusion is justified because of the process that led me to this belief. If so, the following is a perfectly sensible genetic argument:

Sober decided that there are 78 people in the room by drawing the number 78 at random from an urn.

p ════════

It isn't true that there are 78 people in the room.

I have drawn a double line between premiss and conclusion to indicate that the argument is not supposed to be deductively valid. The *p* next to the double line represents the probability that the premiss confers on the conclusion. I claim that *p* is high in this argument.

It is quite true that one cannot *deduce* that a proposition is false just from a description of how someone came to believe it. But I see no reason to think that the context of discovery never provides any evidence at all about whether a belief is true (or plausible). If this is right, we must be careful to distinguish two different formulations of what the genetic fallacy is supposed to involve:

(1) Conclusions about the truth of a proposition cannot be *deduced validly* from premisses that describe how someone came to believe the proposition.

(2) Conclusions about the truth of a proposition cannot be *inferred* from premisses that describe how someone came to believe the proposition.

I think that (1) is true but (2) is false. Inference encompasses more than deductive inference. I conclude that argument (S2) for ethical subjectivism cannot be dismissed simply with the remark that it commits "the genetic fallacy."

The genetic argument concerning how I arrived at my belief about the number of people in the room is convincing. Why? Because *what caused me to reach the belief had nothing whatever to do with whether the belief is true*. When this *independence relation* obtains, the genetic argument shows that the belief is implausible. In contrast, when a *dependence relation* obtains, the description of the belief's genesis can lead to the conclusion that the belief is probably correct.

As an example of how genetic arguments can show that what you believe is probably true, consider my colleague Rebos, who decided that there are 104 people in her philosophy class by carefully counting the people present. I take it that the premiss in the following argument confers a high probability on the conclusion:

Rebos carefully counted the people in her class and consequently believed that 104 people were present.

p ══════════

104 people were present in Rebos's class.

When Rebos did her methodical counting, the thing that caused her to believe that there were 104 people present was *not* independent of how many people actually were there. Because the process of belief formation was influenced in the right way by how many people really were in the room, we are prepared to agree that a description of the context of *discovery* provides a *justification* of the resulting belief.

Let us turn now to the argument for ethical subjectivism summarized by (S2). As the comparison of Sober and Rebos shows, (S2) is incomplete. We need to add some premiss about how the process by which we arrive at our moral beliefs is related to which moral beliefs (if any) are true. Suppose we were to agree with the following thesis:

(A) The processes that determine what moral beliefs people have are entirely independent of which moral statements (if any) are true.

This proposition, if true, would support the following conclusion: *The moral beliefs we currently have are probably untrue.*

The first thing to notice about this conclusion is that it does *not* say that ethical subjectivism is correct. It says that our *current* moral beliefs are probably untrue, not that *all* ethical statements are untrue. Here, we have an important difference between (S2) and the quite legitimate genetic arguments about Sober and Rebos. Clearly, a genetic argument might make plausible the thesis that the ethical statements we happen to believe are untrue. But I do not see how it can show that *no* ethical statements are true.

The next thing to notice about the argument for subjectivism concerns assumption (A). To decide whether (A) is true, we would need to describe (1) the processes that lead people to arrive at their ethical beliefs and (2) the facts about the world, if any, that make those beliefs true or false. We then would have to show that (1) and (2) are entirely independent of each other, as (A) asserts.

Argument (S2) provides a very brief answer to (1)—it cites "evolution" and "socialization." With respect to problem (2), the argument says nothing at all. Of course, if subjectivism were correct, there would be no ethical facts to make ethical beliefs true. But to *assume* that subjectivism is true in the context of this argument begs the question.

Because (S2) says only a little about (1) and nothing at all about (2), I suggest that it is impossible to tell from this argument whether (A) is correct. After all, lots of our beliefs stem either from evolution or from socialization. Mathematical beliefs are of this sort, but that doesn't show that no mathematical statement is true (Kitcher 1985). I conclude that (S2) is a weak argument for ethical subjectivism.

It is not implausible to think that many of our current ethical beliefs are confused. I am inclined to think that morality is one of the last frontiers that human knowledge can aspire to cross. Even harder than the problem of understanding the secrets of the atom, of cosmology, and of genetics is the question of how we ought to lead our lives. This question is harder for us to come to grips with because it is clouded with self-deception: We have a powerful interest in not staring moral issues squarely in the face. No wonder it has taken humanity so long to traverse so modest a distance. Moral beliefs generated by superstition and prejudice probably *are* untrue. Moral beliefs with this sort of pedigree deserve to be undermined by genetic arguments. However, from this critique of some elements of existing morality, one cannot conclude that subjectivism about ethics is correct.

7.5 Models of Cultural Evolution

At present, there is considerable interest and controversy surrounding the application of biological ideas within the social sciences. Sociobiology is the best known of these enterprises. Various philosophical issues raised by sociobiology have been discussed in this chapter. In the present section, I want to discuss a less well-known movement within biology—one that strives to extend evolutionary ideas to social scientific phenomena but not in the way envisioned by sociobiology. Cavalli-Sforza and Feldman (1981) and Boyd and Richerson (1985) have proposed several models of cultural evolution. These authors have distanced themselves from the mistakes they see in sociobiology. In particular, their goal is to describe how cultural traits can evolve for reasons that have nothing to do with the consequences the traits have for survival and reproductive success. In a very real sense, their models describe how mind and culture can play an irreducible and autonomous role in cultural change.

In order to clarify how these models differ from ideas put forward in sociobiology, it will be useful to describe some simple ways in which models of natural selection can differ. As will become clear, I will be using "selection" and other terms more broadly than is customary in evolutionary theorizing.

Given a set of objects that exhibit variation, what will it take for that ensemble to evolve by natural selection? Here, I use "evolve" to mean that the frequency of some characteristic in the population changes. Two ingredients are crucial. The first is that the objects differ with respect to some characteristic that makes a difference in their abilities to survive and reproduce. Then there must be some way to ensure that offspring resemble their parents. The first of these ingredients is differential *fitness;* the second is *heritability* (Lewontin 1970).

In most standard models of natural selection, offspring resemble their parents because a genetic mode of transmission is in place. And traits differ in fitness because some organisms have more babies than others. It may seem odd to say that "having babies" is one way to measure fitness and that passing on genes is one way to ensure heritability, as if there could be others. My reason for saying this will soon become clear.

One way—the most straightforward way—to apply evolutionary ideas to human behavior is to claim that some psychological or cultural characteristic became common in our species by a selection process of the kind just described. This is essentially the pattern of explanation used by Wilson (1975).

The second form that a selection process can take retains the idea that fitness is measured by how many babies an organism produces, but it drops the idea that the relevant phenotypes are genetically transmitted. For example, if characteristics are transmitted because children imitate their parents, a selection process can occur without the mediation of genes.

The incest taboo provides a hypothetical example of how this might happen. Suppose that incest avoidance is advantageous because individuals with the trait have more viable offspring than individuals without it. If offspring *learn* whether to be incest avoiders from their parents, the frequency of the trait in the population may evolve. This could occur without there being any genetic differences between those who avoid incest and those who do not (Colwell and King 1983).

In this second kind of selection model, mind and culture displace one but not the other of the ingredients found in models of the first type. In the first sort of model, a genetic mode of transmission works side by side with a concept of fitness defined in terms of reproductive output—what I have called "having babies." In the second, reproductive output is retained as the measure of fitness, but the genetic mode of transmission is replaced by a psychological one. Learning can provide the requisite heritability just as much as genes.

The third pattern for applying the idea of natural selection abandons both of the ingredients present in the first. The mode of transmission is not genetic, and fitness is not measured by how many babies an organism has. According to this pattern, individuals acquire their ideas because they are exposed to the ideas of their parents, of their peers, and of their parents' generation; transmission patterns may be vertical, horizontal, and oblique. An individual exposed to this mix of ideas need not give them all equal credence. Some may be more attractive than others. If so, the frequency of ideas in the population may evolve. Notice that there is no need for organisms to differ in their survivorship or degree of reproductive success in this case. Some ideas catch on while others become passé. In this third sort of selection model, ideas spread the way a contagion spreads.

Three Types of Selection Model

	heritability	fitness
I	genes	having babies
II	learning	having babies
III	learning	having students

Figure 7.4 Selection processes of type I are standard in discussions of "biological" evolution; those of type III underlie discussions of "cultural" evolution. Those of type II are, so to speak, intermediate.

The *theory of the firm* in economics (discussed in Hirshliefer 1977) is an example of this third type of selection model. Suppose one wishes to explain why businesses behave as profit maximizers. One hypothesis might be that individual managers are rational and economically well informed; they intelligently adjust their behavior to cope with market conditions. Call this the learning hypothesis. An alternative hypothesis is that managers are not especially rational but that inefficient firms go bankrupt and thereby disappear from the market. This second hypothesis posits a selection process of type three. The mode of transmission is not genetic; a business sticks to the same market strategy out of inertia (not because the genes of managers are passed along to their successors). In addition, biological fitness does not play a role. Firms survive differentially, but this does not require individual managers to die or reproduce.

Another example of type three models may be found in some versions of *evolutionary epistemology*. Popper (1973) suggests that scientific theories compete with each other in a struggle for existence. Better theories spread through the population of researchers; inferior ones exit from the scene. Other models in evolutionary epistemology are structured similarly (Toulmin 1972; Campbell 1974; Hull 1988).

The three forms that a selection model can take are summarized in Figure 7.4. "Learning" here should be taken broadly; it doesn't require anything very cognitive but can simply involve imitation. The same goes for "having students"—all that is involved is successful influence mediated by learning.

The parallelism between types I and III is instructive. In type I processes, individuals produce different numbers of babies in virtue of the phenotypes they have (which are transmitted genetically); in type III, individuals produce different numbers of students in virtue of the phenotypes they have (which are transmitted by learning).

Selection models of cultural characteristics that are of either type I or type II can properly be said to provide a "biological" treatment of the traits in question. Models of type III, on the other hand, do not propose biological explanations at all. In type III models, the mode of transmission and the reason for differential survival and replication of ideas may have an entirely autonomous cultural basis.

This threefold division is, of course, consistent with the existence of models that combine two or more of these sorts of process. My taxonomy describes "pure types," so to speak, whereas it is often interesting to consider models in which various pure types are mixed. This is frequently the case in the examples developed by Cavalli-Sforza and Feldman (1981) and by Boyd and Richerson (1985), one of which I'll now describe.

In the nineteenth century, Western societies exhibited an interesting demographic change, one that had three stages. First, oscillations in deathrates due to epidemics and famines became both less frequent and less extreme. Then, overall mortality rates began to decline. The third part of the demographic transition was a dramatic decline in birthrates.

Why did fertility decline? From a narrowly Darwinian point of view, this change is puzzling. A characteristic that *increases* the number of viable and fertile offspring will spread under natural selection, at least when that process is conceptualized by a type I model. Cavalli-Sforza and Feldman are not tempted to appeal to the theory of optimal clutch size developed by Lack (1954), according to which a parent can augment the number of offspring surviving to adulthood by having fewer babies. This Darwinian option is not plausible since women in nineteenth-century Western Europe could have had more viable offspring than they did in fact. People were not caught in the bind that Lack attributed to his birds.

The trait of having fewer children entails a reduction in biological fitness. The trait spread *in spite of* its biological fitness, not *because* of it. In Italy, women changed from having about five children, on average, to having about two. The new trait was far less fit than the old one it displaced.

Cavalli-Sforza and Feldman focus on the problem of explaining how the new custom spread. One possible explanation is that women in all social strata gradually and simultaneously reduced their fertilities. A second possibility is that two dramatically different traits were in competition and that the displacement of one by the other cascaded from higher social classes to lower ones. The first hypothesis, which posits a gradual spread of innovation, says that fertilities declined from 5 to 4.8 to 4.5 and so on, with this process occurring simultaneously across all classes. The second hypothesis says that the trait of having five children competed with the trait of having two and that the novel character was well on its way to displacing the more traditional one among educated people before the same process began among less educated people. There is evidence favoring the second pattern, at least in some parts of Europe.

Cavalli-Sforza and Feldman emphasize that this demographic change could not have taken place if traits were passed down solely from mothers to daughters. This point holds true whether fertility is genetically transmitted or learned. A woman with the new trait will pass it along to fewer offspring than a woman with the old one, if a daughter is influenced just by her mother.

What the process requires is some mixture of horizontal and oblique transmission. That is, a woman's reproductive behavior must be influenced by her peers and by her mother's contemporaries. However, it will not do for a woman to adopt the behavior she finds represented on average in the group that influences her. A woman must find small family size more attractive than large family size even when very few of her peers possess the novel characteristic. In other words, there must be a "transmission bias" in favor of the new trait.

Having a small family was more attractive than having a large one, even though the former trait had a lower Darwinian fitness than the latter. Cavalli-Sforza and Feldman show how the greater attractiveness of small family size can be modeled by ideas drawn from evolutionary theory. However, when these biological ideas are transposed into a cultural setting, one is talking about cultural fitness, not biological fitness. The model they construct of the demographic transition combines two selection processes. When fitness is defined in terms of having babies, there is selection *against* having a small family. When fitness is defined in terms of the psychological attractiveness of an idea, there is selection *favoring* a reduction in family size. Cavalli-Sforza and Feldman show how the cultural process can overwhelm the biological one; given that the trait is sufficiently attractive (and their models have the virtue of giving this idea quantitative meaning), the trait can evolve in spite of its Darwinian disutility.

What are we to make of the research program in which models like this one are developed? Biologists interested in culture are often struck by the absence of viable general theories in the social sciences. All of biology is united by the theory of biological evolution (Section 1.2). Perhaps progress in the social sciences is impeded because there is no general theory of cultural evolution.

The analogies between cultural and genetic change are palpable. And at least some of the disanalogies can be taken into account when the biological models are transposed. For example, we know that biological variation is "undirected"; mutations do not occur because they would be beneficial. In contrast, ideas are *not* invented at random. Individuals often create new ideas—in science, for example—precisely because they would be useful. Another and related disanalogy concerns the genotype/phenotype distinction and the idea that there is no "inheritance of acquired characteristics" (Section 4.4). These principles may have no ready analogs in cultural transmission.

These disanalogies between genetic and cultural change do not show that it is pointless or impossible to write models of cultural evolution that draw on the mathematical resources of population biology. These and other structural differences between biological and cultural evolution can easily be taken into account in models of cultural change.

Another reservation that has been voiced about models of cultural evolution is that they atomize cultural characteristics. Having two children rather than five is a characteristic that is abstracted from a rich and interconnected network of traits. The worry is that by singling out a trait for theoretical treatment, we lose sight of the context that gives that trait cultural meaning.

It is worth realizing that precisely the same question has been raised about biological evolution itself. If you wish to understand the population frequency of sickle-cell anemia, for example, you cannot ignore the fact that the trait is connected with resistance to malaria. In both cultural and biological evolution, it is a mistake to think that each trait evolves independently of all the others. The lesson here is that individual traits should be understood in terms of their relationship to each other.

Although the criticisms I have reviewed so far do not seem very powerful, one rather simple fact about these models suggests that they may be of limited utility in the social sciences. Insofar as these models describe culture, they describe systems of cultural transmission and the evolutionary consequences of such systems. *Given* that the idea of having two children was more attractive than the idea of having five and *given* the horizontal and oblique transmission systems then in place, we can see why the demographic transition took place. But as Cavalli-Sforza and Feldman recognize, their model does not describe *why* educated women in nineteenth-century Italy came to prefer having smaller families, nor *why* patterns adopted in higher classes cascaded down to lower ones. The model describes the *consequences* of an idea's attractiveness, not the *cause* of its being attractive (a distinction introduced in Section 1.6). Historians, on the other hand, will see the real challenge to be the identification of causes.

Models of cultural transmission describe the *quantitative* consequences of systems of cultural influence. Social scientists inevitably make *qualitative* assumptions about the consequences of these systems. If these qualitative assumptions are wrong in important cases and these mistakes actually undermine the plausibility of various historical explanations, social scientists will have reason to take an interest in these models of cultural evolution. But if the qualitative assumptions are correct, historians will have little incentive to take the details of these models into account.

The criticism I have just formulated does not apply to a number of ideas put forward by Boyd and Richerson. They claim that human groups obey a principle of conformist cultural transmission (meaning that individuals adopt traits that are common in their group) and that this learning strategy evolved because of the biological advantages it provided to individuals migrating into new groups. Here we find an effort to describe *and* explain cultural transmission. Whether this two-part proposal is well supported by data is a question I won't try to assess here.

The distinction between source and consequence drawn in connection with models of cultural evolution also applies to some ideas in evolutionary epistemology, including evolutionary models of scientific change. Despite various disanalogies between genes and ideas, the thought that the mix of ideas in a scientific community evolves by a process of "selective-retention" has considerable plausibility (Campbell 1974; Hull 1988; see also Dawkins's 1976 remarks about "memes"). However, the question then arises of what makes one scientific idea "fitter" than another.

Historians of science address this question, though not in this language, when they consider "internalist" and "externalist" explanations of scientific change. Does one idea supplant another because it is better confirmed by observations? Or do scientific ideas come and go because of their ideological utility, their metaphysical palatability, or the power and influence of the people who promulgate them? Clearly, different episodes of scientific change may have different kinds of explanation, and a given change may itself be driven by a plurality of causes. Evolutionary models of scientific change inevitably lead back to these standard problems about *why* scientific ideas change. It seems harmless to agree that fitter theories spread; the question is what makes a theory fitter.

In spite of my criticism of models of cultural evolution, there is something important that these models have achieved. A persistent theme in debates about sociobiology is the relative "importance" that should be accorded to biology and culture. I place the term "importance" in quotation marks because it cries out for clarification. What does it mean to compare the "strength" or "power" of biological and cultural influences?

One virtue of these models of cultural evolution is that they describe culture and biology within a common framework, so that their relative contributions to an outcome are rendered commensurable. What becomes clear in these models is that in comparing the importance of biology and culture, *time is of the essence*. Culture is often a more powerful determiner of change than biological evolution because cultural changes occur *faster*. When biological fitness is calibrated in terms of having babies, its basic temporal unit is the span of a generation. Think how many replication events can occur in that temporal interval when the reproducing entities are ideas that jump from head to head. Ideas spread so fast that they can swamp the slower (and hence weaker) impact of biological natural selection.

This point recapitulates a theme introduced in Section 4.3. In the evolution of biological altruism, group selection pushes in one direction, but individual selection pushes in another. Whether altruism evolves depends on the relative strengths of these selection pressures. But what does "strength" mean in this context? If offspring exactly resemble their parents, the strength of selection is measured by the expected amount of change *per unit time*. If altruism is to evolve, differential survival and reproduction at the group level (i.e., extinction and colonization) must happen *fast enough*.

There is a vague idea about the relation of biology and culture that models of cultural evolution help lay to rest. This is the idea that the science of biology is "deeper" than the social sciences, not just in the sense that it has developed further but in the sense that it investigates more important causes. The inclination is to think that if Darwinian selection favors one trait but cultural influences favor another, the deeper influence of biology must overwhelm the more superficial influence of culture. Cavalli-Sforza and Feldman and Boyd and Richerson deserve credit for showing why this common opinion rests on a confusion.

The conclusion to be drawn here is not that cultural selection *must* overwhelm biological selection when the two conflict but that this *can* happen. Again, the similarity with the conflict between individual and group selection is worth remembering. When two selection processes oppose each other, which "wins" is a contingent matter. The fact that a reduction of family size occurred in nineteenth-century Italy says nothing about what will be true of other traits in other circumstances. The human brain *can* throw a monkey wrench into an adaptationist approach to human behavior. Whether it does so is to be settled on a case-by-case basis.

It is a standard idea in evolutionary theory that an organ will have characteristics that are not part of the causal explanation of why it evolved (Section 4.2). The heart makes noise, but that is not why the heart evolved—it evolved because it pumps blood. Making noise is a *side effect*; it is evolutionary *spin-off* (Section 3.7). We must not lose sight of this distinction when we consider the human mind/brain. Although the organ evolved because of *some* of the traits it has, this should not lead us to expect that *every* behavior produced by the human mind/brain is adaptive. The brain presumably has many side effects; it generates thoughts and feelings that have nothing to do with why it evolved.

Both brains and hearts have features that are adaptations and features that are evolutionary side effects. But to this similarity we must add a fundamental difference. When my heart acquires some characteristic (e.g., a reduced circulation), there is no mechanism in place that causes that feature to spread to other hearts. In contrast, a thought—even one that is neutral or deleterious with respect to my survival and reproduction—is something that may spread beyond the confines of the single brain in which it originates. Brains are linked to each other by networks of mutual influence; it is these networks that allow ideas that occur in one head to influence ideas that occur in others. This is an arrangement that our brains have effected but our hearts have not.

The idea that cultural evolution can swamp biological evolution does not imply that standard processes of biological evolution no longer operate in our species. Individuals still live and die differentially, and differential mortality often has a genetic component. This biological process is not *erased* by the advent of mind and culture: It remains in place but is joined by a second selection process that is made possible by the human mind.

It is quite true that biological evolution produced the brain and that the brain is what causes us to behave as we do. However, it does not follow from this that the brain plays the role of a passive proximate mechanism, simply implementing whatever behaviors happen to confer a Darwinian advantage. Biological selection produced the brain, but the brain has set into motion a powerful process that can counteract the pressures of biological selection. The mind is more than a device for generating the behaviors that biological selection has favored. It is the basis of a selection process of its own, defined by its own measures of fitness and heritability. Natural selection has given birth to a selection process that has floated free.

Suggestions for Further Reading

Caplan (1978) brings together some of the initial salvos in the sociobiology debate, as well as some earlier documents. Kitcher (1985) develops detailed criticisms of what he terms "pop sociobiology" but has positive things to say about other work on the evolution of behavior. Richards (1987) focuses mainly on the history of nineteenth-century evolutionary accounts of mind and behavior. Ruse (1986) argues that evolutionary theory can throw considerable light on traditional philosophical problems about knowledge and values.

References

Alcock, J. (1989): *Animal Behavior: An Evolutionary Approach.* Sinauer.

Alexander, R. (1979): *Darwinism and Human Affairs.* University of Washington Press.

______. (1987): *The Biology of Moral Systems.* Aldine de Gruyter.

Allee, W., Emerson, A., Park, O., Park, T., and Schmidt, K. (1949): *Principles of Animal Ecology.* W. B. Sanders.

Allen, E., et al. (1976): Sociobiology—another biological determinism. Reprinted in A. Caplan (ed.), *The Sociobiology Debate.* New York: Harper & Row, 1978.

Alvarez, L., Alvarez, W., Asaro, F., and Michel, H. (1980): Extraterrestrial cause for the Cretaceous-Tertiary extinction. *Science* 208: 1095–1108.

Aquinas, Thomas (1265): *Summa Theologiae.* In A. Pegis (ed.), *The Basic Writings of St. Thomas Aquinas.* Doubleday. 1955.

Axelrod, R. (1984): *The Evolution of Cooperation.* Basic Books.

Ayala, F. (1974): Introduction. In F. Ayala and T. Dobzhansky (eds.), *Studies in the Philosophy of Biology.* Macmillan.

______. (1987): The biological roots of morality. *Biology and Philosophy* 2: 235–252.

Barash, D. (1979): *The Whisperings Within.* Penguin.

Beatty, J. (1980): Optimal-design models and the strategy of model building in evolutionary biology. *Philosophy of Science* 47: 532–561.

______. (1981): What's wrong with the received view of evolutionary theory? In P. Asquith and R. Giere (eds.), *PSA 1980,* vol. 2, Philosophy of Science Association, 397–426.

______. (1987): On behalf of the semantic view. *Philosophy and Biology* 2: 17–23.

Boorse, C. (1976): Wright on functions. *Philosophical Review* 85: 70–86. Reprinted in Sober (forthcoming-a).

Boyd, R., and Richerson, P. (1985): *Culture and the Evolutionary Process.* University of Chicago Press.

Brandon, R. (1978): Adaptation and evolutionary theory. *Studies in the History and Philosophy of Science* 9: 181–206.

______. (1982): The levels of selection. In P. Asquith and T. Nickles (eds.), *PSA 1,* Philosophy of Science Association, 315–322. Reprinted in Brandon and Burian (1984).

______. (1990): *Organism and Environment.* Princeton University Press.

Brandon, R., and Burian, R. (1984): *Genes, Organisms, and Populations.* MIT Press.

Brockman, J., Grafen, A., and Dawkins, R. (1979): Evolutionarily stable nesting strategy in a digger wasp. *Journal of Theoretical Biology* 77: 473–496.

Bromberger, S. (1966): Why-questions. In R. Colodny (ed.), *Mind and Cosmos.* University of Pittsburgh Press.

Brownmiller, S. (1975): *Against Our Will—Men, Women and Rape*. Bantam.
Burian, R. (1983): Adaptation. In M. Grene (ed.), *Dimensions of Darwinism*. Cambridge University Press.
Cain, A. J. (1989): The perfection of animals. *Biological Journal of the Linnaean Society* 36: 3–29.
Campbell, D. (1974): Evolutionary epistemology. In P. Schilpp (ed.), *The Philosophy of Karl Popper*. Open Court Publishing.
Caplan, A. (ed.) (1978): *The Sociobiology Debate*. Harper & Row.
Cartwright, N. (1979): Causal laws and effective strategies. *Nous* 13: 419–437.
Cassidy, J. (1978): Philosophical aspects of the group selection controversy. *Philosophy of Science* 45: 575–594.
Cavalli-Sforza, L., and Feldman, M. (1981): *Cultural Transmission and Evolution: A Quantitative Approach*. Princeton University Press.
Chorover, S. (1980): *From Genesis to Genocide: The Meaning of Human Nature and the Power of Behavioral Control*. MIT Press.
Clutton-Brock, T., and Harvey, P. (1977): Primate ecology and social organization. *Journal of the Zoological Society of London* 183: 1–39.
Colwell, R., and King, M. (1983): Disentangling genetic and cultural influences on human behavior: problems and prospects. In D. Rajecki (ed.), *Comparing Behavior: Studying Man Studying Animals*. Lawrence Erlbaum Publishers.
Crick, F. (1968): The origin of the genetic code. *Journal of Molecular Biology* 38: 367–379.
Crow, J., and Kimura, M. (1970): *An Introduction to Population Genetics Theory*. Burgess Publishing.
Cummins, R. (1975): Functional analysis. *Journal of Philosophy* 72: 741–764. Reprinted in Sober (forthcoming-a).
Darwin, C. (1859): *On the Origin of Species*. Harvard University Press, 1964.
______. (1871): *The Descent of Man, and Selection in Relation to Sex*. Princeton University Press, 1981.
Dawkins, R. (1976): *The Selfish Gene*. Oxford University Press, 2d ed., 1989.
______. (1979): Twelve misunderstandings of kin selection. *Zeitschrift fur Tierpsychologie* 51: 184–200.
______. (1982a): *The Extended Phenotype*. Freeman.
______. (1982b): Universal Darwinism. In D. Bendall (ed.), *Evolution from Molecules to Men*. Cambridge University Press.
______. (1986): *The Blind Watchmaker*. Longman.
Dequeiroz, K., and Donoghue, M. (1988): Phylogenetic systematics and the species problem. *Cladistics* 4: 317–338.
Descartes, R. (1641): *Meditations on First Philosophy*, translated by Donald Cress. Hackett Publishing. 1979.
Dobzhansky, T. (1971): Evolutionary oscillations in *Drosophila pseudoobscura*. In R. Creed (ed.), *Ecological Genetics and Evolution*. Blackwells.
______. (1973): Nothing in biology makes sense except in the light of evolution. *American Biology Teacher* 35: 125–129.
Doolittle, W., and Sapienza, C. (1980): Selfish genes, the phenotypic paradigm, and genome evolution. *Nature* 284: 601–603.
Edwards, A. (1972): *Likelihood*. Cambridge University Press.
Edwards, A., and Cavalli-Sforza, L. (1963): The reconstruction of evolution. *Annals of Human Genetics* 27: 105.
______. (1964): Reconstruction of evolutionary trees. In V. Heywood and J. McNeill (eds.), *Phenetic and Phylogenetic Classification*. New York Systematics Association Publication 6: 67–76.

Eells, E. (1984): Objective probability theory theory. *Synthese* 57: 387–444.
Ehrlich, P., and Raven, P. (1969): Differentiation of populations. *Science* 165: 1228–1232.
Eldredge, N. (1985): *The Unfinished Synthesis*. Oxford University Press.
Eldredge, N., and Cracraft, J. (1980): *Phylogenetic Patterns and the Evolutionary Process*. Columbia University Press.
Eldredge, N., and Gould, S. (1972): Punctuated equilibria, an alternative to phyletic gradualism. In T. Schopf (ed.), *Models in Paleobiology*. Freeman Cooper.
Ereshefsky, M. (1991a): The semantic approach to evolutionary theory. *Biology and Philosophy* 6: 59–80.
______. (1991b): Species, higher taxa, and the units of evolution. *Philosophy of Science* 58: 84–101.
______. (1992): *The Units of Evolution: Essays on the Nature of Species*. MIT Press.
Farris, S. (1983): The logical basis of phylogenetic analysis. In N. Platnick and V. Funk (eds.), *Advances in Cladistics*, vol. 2. Columbia University Press.
Felsenstein, J. (1978): Cases in which parsimony or compatibility methods will be positively misleading. *Systematic Zoology* 27: 401–410.
______. (1981): A likelihood approach to character weighting and what it tells us about parsimony and compatibility. *Biological Journal of the Linnaean Society* 16: 183–196.
______. (1983): Parsimony methods in systematics—biological and statistical issues. *Annual Review of Ecology and Systematics* 14: 313–333.
______. (1984): The statistical approach to inferring evolutionary trees and what it tells us about parsimony and compatibility. In T. Duncan and T. Stuessy (eds.), *Cladistics: Perspectives on the Reconstruction of Evolutionary History*. Columbia University Press.
Felsenstein, J., and Sober, E. (1986): Likelihood and parsimony—an exchange. *Systematic Zoology* 35: 617–626.
Fisher, R. (1930): *The Genetical Theory of Natural Selection*. Dover Books, 2d ed., 1957.
Ford, E. (1971): *Ecological Genetics*. Chapman and Hall.
Futuyma, D. (1982): *Science on Trial: The Case for Evolution*. Pantheon Books.
______. (1986): *Evolutionary Biology*. Sinauer.
Gaffney, E. (1979): An introduction to the logic of phylogeny reconstruction. In J. Cracraft and N. Eldredge (eds.), *Phylogenetic Analysis and Paleontology*. Columbia University Press.
Ghiselin, M. (1974): A radical solution to the species problem. *Systematic Zoology* 23: 536–544.
______. (1987): Species concepts, individuality, and objectivity. *Biology and Philosophy* 38: 225–242.
Gish, D. (1979): *Evolution? The Fossils Say No!* Creation-Life Publishers.
Gould, S. (1980a): Is a new and general theory of evolution emerging? *Paleobiology* 6: 119–130.
______. (1980b): *The Panda's Thumb*. Norton.
______. (1989): *Wonderful Life*. Norton.
Gould, S., and Lewontin, R. (1979): The spandrels of San Marco and the Panglossian paradigm—a critique of the adaptationist programme. *Proceedings of the Royal Society of London B* 205: 581–598.
Griesmer, J., and Wade, M. (1988): Laboratory models, causal explanations, and group selection. *Biology and Philosophy* 3: 67–96.
Guyot, K. (1987): Specious individuals. *Philosophica* 37: 101–126.
Haldane, J. (1927): A mathematical theory of natural and artificial selection, V: Selection and mutation. *Proceedings of the Cambridge Philosophical Society* 23: 838–844.
______. (1932): *The Causes of Evolution*. Cornell University Press.

Hamilton, W. (1964): The genetical evolution of social behavior. *Journal of Theoretical Biology* 7: 1–52.
______. (1967): Extraordinary sex ratios. *Science* 156: 477–488.
Harper, J. (1977): *Population Biology of Plants*. Academic Press.
Hempel, C. (1965a): *Aspects of Scientific Explanation and Other Essays in the Philosophy of Science*. Free Press.
______. (1965b): *Philosophy of Natural Sciences*. Prentice-Hall.
Hennig, W. (1965): Phylogenetic systematics. *Annual Review of Entomology* 10: 97–116.
______. (1966): *Phylogenetic systematics*. University of Illinois Press. Revision and translation of Hennig's 1950 *Grundzuge einer Theorie der phylogenetishen Systematik*.
Hesse, M. (1969): Simplicity. In P. Edwards (ed.), *The Encyclopedia of Philosophy*. Macmillan.
Hirshliefer, J. (1977): Economics from a biological viewpoint. *Journal of Law and Economics* 1: 1–52.
Horan, B. (1989): Functional explanations in sociobiology. *Biology and Philosophy* 4: 131–158.
Hull, D. (1965): The effect of essentialism on taxonomy—2000 years of stasis. *British Journal for the Philosophy of Science* 15: 314–326; 16: 1–18.
______. (1970): Contemporary systematic philosophies. *Annual Review of Ecology and Systematics* 1: 19–54.
______. (1974): *Philosophy of Biological Sciences*. Prentice-Hall.
______. (1976): Informal aspects of theory reduction. In R. S. Cohen et al. (eds.), *PSA 1974*. Dordrecht, 653–670.
______. (1978): A matter of individuality. *Philosophy of Science* 45: 335–360. Reprinted in Sober (forthcoming-a).
______. (1980): Individuality and selection. *Annual Review of Ecology and Systematics* 11: 311–332.
______. (1988): *Science as a Process*. University of Chicago Press.
Hume, D. (1779): *Dialogues Concerning Natural Religion.* Thomas Nelson and Sons. 1947.
______. (1739): *Treatise of Human Nature.* Oxford University Press. 1968.
Jablonski, D. (1984): Keeping time with mass extinctions. *Paleobiology* 10: 139–145.
Jacob, F. (1977): Evolution and tinkering. *Science* 196: 1161–1166.
Janzen, D. (1977): What are dandelions and aphids? *American Naturalist* 111: 586–589.
Johnson, L. (1970): Rainbow's end—the quest for an optimal taxonomy. *Systematic Zoology* 19: 203–239.
Joseph, G. (1980): The many sciences and the one world. *Journal of Philosophy* 77: 773–790.
Kettlewell, H. (1973): *The Evolution of Melanism*. Oxford University Press.
Kim, J. (1978): Supervenience and nomological incommensurables. *American Philosophical Quarterly* 15: 149–156.
Kimura, M. (1983): *The Neutral Theory of Molecular Evolution*. Cambridge University Press.
Kitcher, P. (1982a): *Abusing Science: The Case Against Creationism*. MIT Press.
______. (1982b): Genes. *British Journal for the Philosophy of Science* 33: 337–359.
______. (1984): Species. *Philosophy of Science* 51: 308–333.
______. (1985): *Vaulting Ambition: Sociobiology and the Quest for Human Nature*. MIT Press.
Kolmogorov, A. (1933): *Foundations of the Theory of Probability*. Chelsea.
Krebs, J., and Davies, N. (1981): *An Introduction to Behavioural Ecology*. Sinauer.
Lack, D. (1954): *The Optimal Regulation of Animal Numbers*. Oxford University Press.

Lakatos, I. (1978): Falsification and the methodology of scientific research programmes. In I. Lakatos, *The Methodology of Scientific Research Programmes: Philosophical Papers*, vol. 1. Cambridge University Press.
Lamarck, J. (1809): *Zoological Philosophy*. University of Chicago Press, 1984.
Laudan, L. (1977): *Progress and Its Problems*. University of California Press.
Levins, R. (1966): The strategy of model building in population biology. *American Scientist* 54: 421–431.
Lewontin, R. (1970): The units of selection. *Annual Review of Ecology and Systematics* 1: 1–14.
______. (1974): The analysis of variance and the analysis of causes. *American Journal of Human Genetics* 26: 400–411.
______. (1978): Adaptation. *Scientific American* 239: 156–169.
Lewontin, R., and Dunn, L. (1960): The evolutionary dynamics of a polymorphism in the house mouse. *Genetics* 45: 705–722.
Lewontin, R., Rose, S., and Kamin, L. (1984): *Not in Our Genes: Biology, Ideology, and Human Nature*. Pantheon.
Lloyd, E. (1988): *The Structure and Confirmation of Evolutionary Theory*. Greenwood.
Lorenz, K. (1966): *On Aggression*. Methuen.
Maddison, W., Donoghue, M., and Maddison, D. (1984): Outgroup analysis and parsimony. *Systematic Zoology* 33: 83–103.
Malinowski, B. (1922): *Argonauts of the Western Pacific*. E. P. Dutton, 1961.
Maynard Smith, J. (1977): *The Theory of Evolution*. Penguin.
______. (1978a): *The Evolution of Sex*. Cambridge University Press.
______. (1978b): Optimization theory in evolution. *Annual Review of Ecology and Systematics* 9: 31–56. Reprinted in Sober (forthcoming-a).
______. (1982): *Evolution and the Theory of Games*. Cambridge University Press.
______. (1989): *Evolutionary Genetics*. Oxford University Press.
Maynard Smith, J., and Price, G. (1973): The logic of animal conflict. *Nature* 246: 15–18.
Mayr, E. (1942): *Systematics and the Origin of Species*. Columbia University Press.
______. (1961): Cause and effect in biology. *Science* 134: 1501–1506. Reprinted in E. Mayr, *Towards a New Philosophy of Biology*. Harvard University Press, 1988.
______. (1963): *Animal Species and Evolution*. Harvard University Press.
______. (1974): Teleological and teleonomic: a new analysis. *Boston Studies in the Philosophy of Science* 14: 91–117. Reprinted as "The multiple meanings of teleological" in E. Mayr, *Towards a New Philosophy of Biology*. Harvard University Press, 1988.
______. (1981): Biological classification—towards a synthesis of opposing methodologies. *Science* 214: 510–516.
Mayr, E., and Ashlock, P. (1991): *Principles of Systematic Zoology*. McGraw-Hill.
Michod, R., and Sanderson, M. (1985): Behavioral structure and the evolution of cooperation. In J. Greenwood and M. Slatkin (eds.), *Evolution: Essays in Honor of John Maynard Smith*. Cambridge University Press.
Mills, S., and Beatty, J. (1979): The propensity interpretation of fitness. *Philosophy of Science* 46: 263–288. Reprinted in Sober (forthcoming-a).
Mishler, B., and Donoghue, M. (1982): Species concepts—a case for pluralism. *Systematic Zoology* 31: 491–503.
Mitchell, S. (1987): Competing units of selection? *Philosophy of Science* 54: 351–367.
Mitchell, W., and Valone, T. (1990): The optimization research program: studying adaptations by their function. *Quarterly Review of Biology* 65: 43–52.
Moore, G. (1903): *Principia Ethica*. Cambridge University Press. 1968.
Morris, H. (1974): *Scientific Creationism*. Creation-Life Publishers.

Nelson, G. (1979): Cladistic analysis and synthesis: principles and definitions, with a historical note on Adanson's *Familles des Plantes* (1763–1764). *Systematic Zoology* 28: 1–21.

Nelson, G., and Platnick, N. (1981): *Systematics and Biogeography*. Columbia University Press.

Neyman, J. (1950): *First Course in Probability and Statistics*. Henry Holt.

Numbers, R. (1992): *The Creationists*. Knopf.

O'Hara, R. (forthcoming): Systematic generalization, historical fate, and the species problem. *Systematic Biology*.

Orgel, L., and Crick, F. (1980): Selfish DNA: the ultimate parasite. *Nature* 284: 604–607.

Orzack, S., and Sober, E. (forthcoming): Optimality models and the long-run test of adaptationism. *American Naturalist*.

Ospovat, D. (1981): *The Development of Darwin's Theory*. Cambridge University Press.

Oster, G., and Wilson, E. (1978): *Caste and Ecology in the Social Instincts*. Princeton University Press.

Paley, W. (1805): *Natural Theology*. Rivington.

Parker, G. (1978): Search for mates. In J. Krebs and N. Davies (eds.), *Behavioral Ecology: An Evolutionary Approach*. Blackwells.

Parker, G., and Maynard Smith, J. (1990): Optimality theory in evolutionary biology. *Nature* 348: 27–33.

Paterson, H. (1985): The recognition concept of species. In E. Vrba (ed.), *Species and Speciation*. Transvaal Museum Monograph 4: 21–29.

Patterson, C. (1980): Cladistics. *Biologist* 27: 234–240. Reprinted in J. Maynard Smith (ed.), *Evolution Now*. Macmillan, 1982.

______. (1988): The impact of evolutionary theories on systematics. In D. Hawksworth (ed.), *Prospects in Systematics*. Clarendon Press.

Platnick, N. (1979): Philosophy and the transformation of cladistics. *Systematic Zoology* 28: 537–546.

Popper, K. (1959): *The Logic of Scientific Discovery*. Hutchinson.

______. (1963): *Conjectures and Refutations*. Harper.

______. (1973): *Objective Knowledge*. Oxford University Press.

______. (1974): Darwinism as a metaphysical research program. In P. Schilpp (ed.), *The Philosophy of Karl Popper*. Open Court.

Putnam, H. (1975): Philosophy and our mental life. In *Mind, Language, and Reality*. Cambridge University Press.

Quine, W. (1952): Two dogmas of empiricism. In W. Quine, *From a Logical Point of View*. Harvard University Press.

______. (1960): *Word and Object*. MIT Press.

______. (1966): On simple theories of a complex world. In W. Quine, *The Ways of Paradox and Other Essays*. Harvard University Press.

Rawls, J. (1970): *A Theory of Justice*. Harvard University Press.

Reichenbach, H. (1956): *The Direction of Time*. University of California Press.

Richards, R. (1987): *Darwin and the Emergence of Evolutionary Theories of Mind and Behavior*. University of Chicago Press.

Ridley, M. (1986): *Evolution and Classification: The Reformation of Cladism*. Longman.

Rose, M., and Charlesworth, B. (1981): Genetics of life history in *Drosophila melanogaster*. *Genetics* 97: 173–196.

Rosenberg, A. (1978): The supervenience of biological concepts. *Philosophy of Science* 45: 368–386.

______. (1983): Fitness. *Journal of Philosophy* 80: 457–473.

______. (1985): *The Structure of Biological Science*. Cambridge University Press.

Rosenberg, A., and Williams, M. (1985): Fitness in fact and fiction. *Journal of Philosophy* 82: 738–749.

Ruse, M. (1973): *The Philosophy of Biology*. Hutchinson.

______. (1980): Charles Darwin on group selection. *Annals of Science* 37: 615–630.

______. (1986): *Taking Darwin Seriously*. Blackwells.

______. (1988a): *But Is It Science?* Prometheus Books.

______. (1988b): *Philosophy of Biology Today*. State University of New York Press.

Ruse, M., and Wilson, E. (1986): Moral philosophy as applied science. *Philosophy* 61: 173–192.

Salmon, W. (1984): *Scientific Explanation and the Causal Structure of the World*. Princeton University Press.

Schaffner, K. (1976): Reduction in biology—problems and prospects. In R. S. Cohen *et al.* (eds.), *PSA 1974*. Dordrecht, 613–632.

Shoemaker, S. (1984): *Identity, Cause, and Mind*. Cambridge University Press.

Simpson, G. (1961): *The Principles of Animal Taxonomy*. Columbia University Press.

Skyrms, B. (1980): *Causal Necessity*. Yale University Press.

Smart, J. (1963): *Philosophy and Scientific Realism*. Routledge.

Sneath, P., and Sokal, R. (1973): *Numerical Taxonomy*. W. H. Freeman.

Sober, E. (1981): Holism, individualism, and units of selection. In P. Asquith and R. Giere (eds.), *PSA 1980*, vol. 2, Philosophy of Science Association, 93–121.

______. (1984a): Fact, fiction, and fitness. *Journal of Philosophy* 81: 372–384.

______. (1984b): *The Nature of Selection*. MIT Press.

______. (1984c): Sets, species, and natural kinds—a reply to Philip Kitcher's "Species." *Philosophy of Science* 51: 334–341.

______. (1987a): Does "fitness" fit the facts? *Journal of Philosophy* 84: 220–223.

______. (1987b): What is adaptationism? In J. Dupré (ed.), *The Latest on the Best*. MIT Press.

______. (1988): *Reconstructing the Past: Parsimony, Evolution, and Inference*. MIT Press.

______. (1990a): Let's razor Ockham's razor. In D. Knowles (ed.), *Explanation and its Limits*. Cambridge University Press.

______. (1990b): The poverty of pluralism—a reply to Sterelny and Kitcher. *Journal of Philosophy* 87: 151–157.

______. (1992a): Organisms, individuals, and the units of Selection. In F. Tauber (ed.), *Organism and the Origins of Self*. Kluwer.

______. (1992b): Screening-off and the units of selection. *Philosophy of Science* 59: 142–152.

______. (forthcoming-a): *Conceptual Issues in Evolutionary Biology*. MIT Press, 2d ed.

______. (forthcoming-b): Evolutionary altruism, psychological egoism, and morality: disentangling the phenotypes. In M. Nitecki (ed.), *Evolutionary Ethics*. SUNY Press.

Sober, E., and Lewontin, R. (1982): Artifact, cause, and genic selection. *Philosophy of Science* 47: 157–180.

Sokal, R., and Crovello, T. (1970): The biological species concept—a critical evaluation. *American Naturalist* 104: 127–153.

Stanley, S. (1979): *Macroevolution: Pattern and Process*. W. H. Freeman.

Sterelny, K., and Kitcher, P. (1988): The return of the gene. *Journal of Philosophy* 85: 339–361.

Templeton, A. (1989): The meaning of species and speciation—a genetic perspective. In D. Otte and J. Endler (eds.), *Speciation and Its Consequences*. Sinauer.

Thompson, P. (1988): *The Structure of Biological Theories*. SUNY Press.

Toulmin, S. (1972): *Human Understanding*. Princeton University Press.

Trivers, R. (1972): The evolution of reciprocal altruism. *Quarterly Review of Biology* 46: 35–57.

______. (1985): *Social Evolution*. Benjamin Cummings.

Van Valen, L. (1976): Ecological species, multispecies, and oaks. *Taxon* 25: 233–239.

Vrba, E. (1980): Evolution, species, and fossils: how does life evolve? *South African Journal of Science* 76: 61–84.

Wade, M. (1978): A critical review of models of group selection. *Quarterly Review of Biology* 53: 101–114.

Waters, K. (1986): Models of natural selection: from Darwin to Dawkins. Doctoral dissertation, Indiana University.

______. (1991): Tempered realism about the forces of selection. *Philosophy of Science* 58: 553–573.

Whitehouse, H. (1973): *Towards an Understanding of the Mechanism of Heredity*. St. Martin's.

Wiley, E. (1981): *Phylogenetics: The Theory and Practice of Phylogenetic Systematics*. John Wiley.

Williams, G. (1966): *Adaptation and Natural Selection*. Princeton University Press.

______. (1985): A defense of reductionism in evolutionary biology. In R. Dawkins and M. Ridley (eds.), *Oxford Surveys in Evolutionary Biology*, vol. 2. Oxford University Press.

Williams, M. (1973): The logical basis of natural selection and other evolutionary controversies. In M. Bunge (ed.), *The Methodological Unity of Science*. D. Reidel.

Wilson, D. (1980): *The Natural Selection of Populations and Communities*. Benjamin Cummings.

______. (1990): Weak altruism, strong group selection. *Oikos* 59: 135–140.

Wilson, D., and Dugatkin, L. (1991): Nepotism vs. tit-for-tat or, why should you be nice to your rotten brother? *Evolutionary Ecology* 5: 291–299.

Wilson, D., and Sober, E. (1989): Reviving the superorganism. *Journal of Theoretical Biology* 136: 337–356.

Wilson, E. (1975): *Sociobiology: The New Synthesis*. Harvard University Press.

______. (1978): *On Human Nature*. Harvard University Press.

______. (1980): Comparative social theory. In S. McMurrin (ed.), *The Tanner Lectures on Human Values*. University of Utah Press.

Wimsatt, W. (1979): Reduction and reductionism. In P. Asquith and H. Kyburg (eds.), *Current Research in Philosophy of Science*. Philosophy of Science Association, 352–377.

______. (1980): Reductionistic research strategies and their biases in the units of selection controversy. In T. Nickles (ed.), *Scientific Discovery*, vol. 2. D. Reidel.

Wright, L. (1976): Functions. *Philosophical Review* 85: 70–86. Reprinted in Sober (forthcoming-a).

Wright, S. (1931): Evolution in Mendelian populations. *Genetics* 16: 97–159.

______. (1945): *Tempo and Mode in Evolution*: A critical review. *Ecology* 26: 415–419.

______. (1968): *Evolution and the Genetics of Populations 1*. University of Chicago Press.

Wynne-Edwards, V. (1962): *Animal Dispersion in Relation to Social Behavior*. Oliver and Boyd.

About the Book and Author

Perhaps because of its implications for our understanding of the kind of beings we are, recent philosophy of biology has seen what may be the most dramatic work in the philosophies of the "special" sciences. This drama has centered on evolutionary theory, and in this new textbook, Elliott Sober introduces the reader to the most important of these developments.

With a rare combination of technical sophistication and clarity of expression, Sober engages both the higher level of theory and the direct implications for such controversial issues as creationism, teleology, nature versus nurture, and sociobiology. Above all, the reader will gain from this book a firm grasp of the structure of evolutionary theory, the evidence for it, and the scope of its explanatory significance.

Elliott Sober is Hans Reichenbach Professor of Philosophy at the University of Wisconsin–Madison. He is editor of *Conceptual Issues in Evolutionary Biology: An Anthology* and author of *Simplicity* and *The Nature of Selection: Evolutionary Theory in Philosophical Focus* as well as many papers on the philosophy of science and of biology. In 1991 he was awarded the Lakatos Award for an outstanding contribution to the philosophy of science for his book *Reconstructing the Past: Parsimony, Evolution, and Inference.*

Index